URBAN POLICY
UNDER
CAPITALISM

Volume 22, URBAN AFFAIRS ANNUAL REVIEWS

URBAN POLICY
UNDER
CAPITALISM

Edited by
NORMAN I. FAINSTEIN
and
SUSAN S. FAINSTEIN

Volume 22, URBAN AFFAIRS ANNUAL REVIEWS

SAGE PUBLICATIONS
Beverly Hills / London / New Delhi

For information address:

SAGE Publications, Inc.
275 South Beverly Drive
Beverly Hills, California 90212

SAGE Publications India Pvt. Ltd.
C-236 Defence Colony
New Delhi 110 024, India

SAGE Publications Ltd
28 Banner Street
London EC1Y 8QE, England

Printed in the United States of America

Library of Congress Cataloging in Publication Data

Main entry under title:

Urban policy under capitalism.

    (Urban affairs annual review ; v. 22)
    Includes bibliographical references.
    1. Urban policy—Addresses, essays, lectures.
2. Economic policy—Addresses, essays, lectures.
3. Social policy—Addresses, essays, lectures.
I. Fainstein, Norman I.   II. Fainstein, Susan S.
III. Series.
HT108.U7   vol. 22   [HT151]   307.7′6s   81-23185
ISBN 0-8039-1797-X   [307.7′6]   AACR2
ISBN 0-8039-1798-8 (pbk.)

SECOND PRINTING, 1983

CONTENTS

Part I

Introduction

1

Restoration and Struggle:
Urban Policy and Social Forces

NORMAN I. FAINSTEIN
SUSAN S. FAINSTEIN

□ THIS COLLECTION OF ESSAYS arises out of the relatively new scholarly tradition of urban political economy (see Harloe, 1977; Fainstein and Fainstein, 1979; and the chapters in this volume by Walton and Rich). The intellectual contribution of this tradition is complex and multidimensional; two aspects of its break with both orthodox Marxism and earlier liberal urban analysis provide the starting point for the authors here. First, even though they trace the roots of urban development to the capitalist mode of production, the authors identify indeterminacy as a consequence of the mediating nature of culture, historic compromises, and contemporary struggle. General tendencies are always confronted by particular circumstances, making specific urban outcomes, in Castells' (1977) term, "problematic." Second, the contributors avoid a prescriptive approach to public policymaking. They assume that policy results from the interaction of economic forces and political interventions. Such a stance does not imply that state policy is meaningless as a factor in social well-being or irrational as the manifestation of particular economic interests; only that the search for an "optimal" policy is both fruitless and ideologically mystifying. Therefore, the study of urban policy requires the analysis of the basis of state action in class, group, and geographical domination and conflict.

There are several reasons for looking specifically at the urban dimensions of public policy, even in the context of a holistic view of policy as rooted in general social forces. Within the development of post-World War II capitalism, the national state has assumed increasing importance in directing capital investment and providing for social welfare. But the key functions of urban space in structuring capital accumulation and determining the quality of everyday life have caused political actors and scholars to define many general issues within the framework of urban analysis. Subnational government, either as agent of the national state or as allocator of borrowed capital, has become a crucial force in supporting and shaping further growth. Social expenditures on transportation, housing, commercial and industrial development, and urban renewal have molded urban form in both the first and third worlds. Within the spatial system, state intervention concerned with land use and infrastructure (i.e., urban development) is critical to capital directly, because increasing proportions of capital investment flow into real estate, and, equally important, because capitalist production depends on agglomeration economies and transportation technology. Indirectly, urban conditions affect the flow of investment within and among cities, with important consequences for the profitability of various economic sectors.

The use of urban space as a vehicle for accumulation, however, clashes with its second principal social function as the territory in which the population resides, consumes goods and services, and acts politically (i.e., as the location of social reproduction). Within this category of analysis, social welfare becomes also a fundamental object of urban policy. While social conflict in the workplace has, during the last two decades, assumed an increasingly ritualized form, urban protest has been volatile, difficult to control, symbolically powerful, and sometimes effective in diverting capitalist state and economy from their accumulation objectives. The function of urban areas as residential locations for the mass of the population, their still rapid growth in the less developed world, competition between different consumption and cultural groups over domination of territory, conflicting interests in the use of the economic surplus for infrastructural investment or social programs — all block the functioning of the local state and territory for exclusively accumulative purposes. The nature of the struggle varies from place to place within and between countries — in the United States between downtown and neighborhoods, landlord and tenant; within neighborhoods between cultural, ethnic, and income groups. In Europe, with its much greater public role in housing and land markets, conflict has most recently become evident in the squatters' movements, but expresses itself also in the capture of local governments by left parties, the constant tension between upper- and working-class claims to territory, and the cultural and social demands of mobilized groups. The less developed countries, dependent on outside capitalists, demonstrate different relationships between state and class, but nevertheless display similar conflicts over territorial rights and investment demands.

The increasingly collective nature of modern capitalism has produced an enlargement of the state's role as mediator among fractions of capital and legitimator of the distribution of benefits within society. The consequence of state expansion has been fiscal crisis (O'Connor, 1973). As the general woes of the capitalist economies have increased in the present period of inflation, slow or nonexistent growth, and rising unemployment, the fiscal crisis of the state has merged with the larger economic crisis to provide a rationale for reducing the size of the state sector in a number of countries. We are, therefore, in the United States and in some Western European states, at a point where powerful forces are seeking to restore the status quo ante of a less prominent role for government in the realm of social reproduction. Whether conservative restoration will succeed in transforming the state into a less active determinant of social arrangements remains to be seen. But there is agreement among the authors in this book that we are in a new stage of urban development, shaped by general, and in fact worldwide political-economic tendencies. Understanding the current stage requires comparative examination of the particular manifestations of urban development and conflict in various countries and localities along with an overall grasp of the macropolitical economy. The chapters herein represent an effort to sort out the determinative effects of economic conditions and the interdeterminate consequences of specific political and cultural mobilizations. In this introduction we sketch some of the theoretical relationships connecting state intervention with social cleavages. We then indicate the ways in which the essays specify various dimensions of these relationships within particular social contexts.

CAPITAL, CONFLICT, AND STATE PROGRAMS

Capitalists require state intervention in the urban system. Through infrastructural investment and public services, the state provides social capital; through coercion, propaganda, and social service expenditures, the state maintains social control. But the content of public policy is not directly specified by general functional necessities, and for two main reasons. First, the institutional separation of the state from capital compels the state to mediate among classes, with liberal-democratic regimes forced to respond to working-class political power, upon occasion to votes instead of money. Second, the interests of capital are divided. The market destroys value as well as creating it, harming fractions of capital tied to declining functions, industries, or territorial units. Taken together, these factors create an indeterminacy which the state experiences in a recurring conundrum: Should government policy and programs reinforce market tendencies, or should they seek to deflect and possibly negate them? In practice, the answer will depend on the tendencies in question and on the balance of class forces experienced within a particular regime.

The state under capitalism is expected by business leaders to mitigate urban problems which are consequences of their own actions. State solutions, however, must not impede the accumulation process by burdening profits with taxes, and they must not threaten directly the character of control over production to which a national business class has become accustomed. Consider, for example, the United States and Britain, whose regimes are faced with the sectoral/spatial restructuring of advanced capitalist economies underway since the mid-seventies. The Reagan approach has so far been to allow capital to move of its own accord and to expect labor to do likewise. At the same time, the Reagan retrenchment of the welfare state may produce a large negative multiplier in the growth sector of the American economy — services. Thatcher, who has not learned supply-side economics, and who runs a more interventionist state confronted with more dependent capitalists, seems intent upon closing Britain's old plant and replacing it with new, presumably more productive, manufacturing. She does not expect labor to move from the old inner cities (where mass unemployment is the consequence), but is unwilling to control the flow of British capital out of manufacturing and out of the country. In both cases, the regime is torn between laissez-faire and intervention, as well as between its desire to serve capital as a whole and its unwillingness to limit the prerogatives of the particular capitalists who comprise its constituency.

Among all the advanced capitalist nations, patterns of urbanism and acute urban problems (over-development, abandonment, social movements) have resulted from the restructuring of the economy. When capital in the West has flowed out of manufacturing, it has frequently been into land and the built environment. North American and Western European real estate has also often been the final resting place of petrodollars. Overinvestment in real estate amounts to speculation which remains safe so long as demand continues high in spite of escalating prices. The supervalorization of real-estate investment — most of which contributes little to actual production or productivity — makes it even more attractive to investors as a hedge against further inflation.

While the consequences of capital flows for urbanization have been enormous, capitalists have no collective interest in a particular urban pattern. Various fractions of capital do, however, have interests tied to the rise or fall of cities and regions. As a result, there is a complex patterning of business interests around the state. At each level of the state (depending on political institutional arrangements), business interests are likely to be divided between those with and those without a major stake in urban development. Of those with such stakes, interests may differ sharply. For example, in the third world, some capitalists are tied entirely to the export sector, while others depend heavily on local investment of the surplus; in developed nations we see numerous cases of conflict within the business class over subsidies to urban growth and welfare — whether government should support urban programs or divert resources to tax reductions or encouragement of agricultural and extractive enterprises.

While the business class experiences internal "fault lines" between different sectoral and geographical interests, it nevertheless shares strong common objectives as it faces the state. These derive from the main cleavages of the entire capitalist social formation, the division between classes, and the division between state and economy. With regard to the former, the struggle, simply put, is between profits and wages — between capital and labor over shares of surplus value, and more generally, over the definition of the social relations of production. As capital relates to the state, the interest of the upper class is in state programs from which it is the prime beneficiary, while others bear the financing burden. In addition, the business class has an equally strong interest in maintaining private control over profitable activities and in itself determining the social relations of production (including investment, division of labor, pay rates).[1]

These considerations may be formalized in a way which helps us to see the class interests and political behavior of capital. Figure 1.1 is intended to be suggestive of the primary dimensions which define the interests of national capital *as a whole class* (thus sectoral and geographical axes are omitted) toward state activity. The vertical dimension of *net income effect* reflects the calculation of who receives the benefits of state programs versus who finances their costs. The hori-

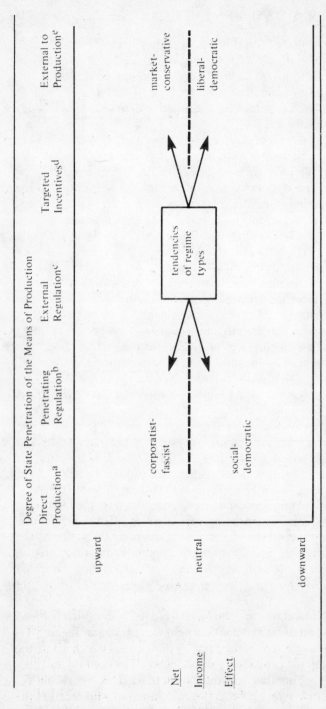

Figure 1.1 Class Interests in State Program Initiatives

a. Public services, banking, utilities, transport, energy, manufacturing.
b. Organization of production, wage structure, union power, investment decisions, personnel recruitment.
c. Product safety, building codes, zoning restrictions, environmental pollution regs.
d. Free trade zones, urban renewal land write-downs, subsidies for hiring unemployed.
e. Cash transfers to firms and individuals—government purchases, tax reductions, welfare payments.

zontal dimension reflects the degree of *state penetration of the means of production*. While this is not the place to elaborate the framework, several points deserve comment.

First, capital has an objective interest *both* in maximizing its share of net state expenditures and in maximizing private control over all aspects of production and investment; hence the two dimensions, one for material distribution, the other for power. Second, while the dimensions are conceptually distinct, they are empirically related to the degree that income effect may be an outcome of program progress. So, to some extent capitalists object to loss of control because they perceive an immediate economic cost to themselves. Nevertheless, the class as a whole reacts (through its leading political organizations) by considering both dimensions. Thus, a program with a moderate "downward" income effect which lies entirely external to production (e.g., family allowances) may be more acceptable than state investment planning controls aimed at enhancing accumulation. Third, the analytic schema is most helpful in understanding the interests of capital at the margin; we ask, in effect, what new programs capital will try to enact or resist, what old ones it will try to dismantle or strengthen. Finally, it is worth noting that while capital *thinks* control and income effect, it *talks* regulation and economic growth.[1]

The actual character of state policy affecting urbanism will ultimately depend upon political factors which we will enumerate here (but which will be described in detail in the articles to follow). Capital acts as a whole only through its political organizations, whether business associations, political parties, or parts of the state administrative apparatus. Accordingly, the interests of capital which become expressed in "urban" policy will depend not only on the sectoral and geographical cleavages we discussed earlier, but especially on the particular elements of the class that have access to the regime in power. The character of the regime, in turn, depends upon the ideology of the leaders of the governing coalition and of the political parties which bring them to power.

Thus, in Figure 1.1 we also suggest the tendencies of four general regime types with regard to public policy. In two of the types we assume that capital is in a hegemonic position; these we refer to as *market-conservative* and *corporatist-fascist*. The *liberal-democratic* regime incorporates working-class interests through the dominance in the governing coalition of a conservative party of the left (e.g., British Labor the last time it ruled, the German Social Democrats, the wing of the U.S. Democrats represented by Johnson). The *social democratic* regime reflects the domination of a leftist working-class party within the capitalist state. Such a regime may have come to power recently in France, and might yet in Britain and Italy. Even if a leftist working-class party does not itself govern, it may pressure the regime to introduce policies which both redistribute income "downward" and penetrate the means of production. Obviously, liberal-democratic regimes are most susceptible to such pressure. Moreover, all regimes must respond to urban social movements, protest, and disorder. The manner in which they respond depends on the balance of class forces they incorporate. For these reasons, the programs of the state under capitalism are always a product of the functional needs of the economy and organization of capitalist interests on the one hand, but of political leadership and class struggle on the other.

As a final introductory note, and before a discussion of the essays themselves, it is important to recapitulate what we mean by *urban* policy. We do not employ the nominal designations by political re-gimes of what is "urban policy." For one thing, "urban policy" is a political symbol with different meanings in different countries and times. For another, it may nominally fade from existence, as seems to be the case with the Reagan administration. But most important, there is no necessary correspondence between programs intended to affect urbanism and those which do, regardless of their objectives and definitions.

What is urban policy then? It is state activity which affects ur-banism. By urbanism we mean the distribution of investment and consumption activities in real space, the character and form of the built environment, and the distribution of population groupings in relation to both. In abstract terms, it follows that *urban policy is state activity affecting the use of space and the built environment relative to the process of accumulation and the social occupation of space relative to the distribution of consumption opportunities*. In practical terms, then, economic policy and social policy are as much urban policy (defined by their urban impacts) as, say, land-use controls or urban redevelopment. In fact, as several authors argue below, urban policy is increasingly being made under some other name.

ANALYTIC APPROACHES OF THE ESSAYS

Fundamental to the political economy approach to urban de-velopment is a rejection of disciplinary boundaries and isolated levels of analysis. Dichotomies between inputs and outputs, politics and policies, economy and society, nation and city (even world and city) disappear. The absence of these familiar categories, however, creates difficult methodological problems for scholars and trying organizational ones for editors of collections such as this. The meaning of the terms "urban" and "policy" tend to melt away as authors seek to root them, on the one hand, in large-scale political and economic tendencies; and on the other, to trace their manifestations in specific cases. Macroanalysis of urban phenomena requires understanding of very broad issues, such as the "recapitalization of capital," the place of peripheral states within the world system, national integration in the third world. and class conflict under advanced capitalism. Specificity can mean both a general examination of Prime Minister Thatcher's social policy and an analysis of San Francisco's gay politics.

We are struck by certain common themes that run through most of the chapters, whatever their particular focus. First, the effort to identify specific outcomes of general historic tendencies leads naturally to comparative analysis. Only this mode can identify what phenomena are unique to particular social formations and what are general trends in modern capitalism. Second, this approach by assuming a wide scope of variability under contemporary capitalism, tends to emphasize the importance of cultural and ideological definitions in channeling governmental and social action. A Weberian construction is introduced. Although ideas do not make history, they do act as "switchmen, determin[ing] the tracks along which action has been pushed by the dynamic of interest" (Gerth and Mills, 1964: 280).

Third, common to all the chapters is an attempt to tie theory to practice, to formulate theoretical propositions inductively from empirical observation rather than deductive analysis. This effort, characteristic of writings by students of urbanism, deviates from the more philosophical method of European critical theory (Anderson, 1979). As well as reflecting a particular theory of knowledge, it arises from a commitment by the authors to the use of knowledge as a basis for action. A number of chapters conclude with paraphrases of Marx's exhortation that the problem is not to interpret the world differently, but to change it (Marx, 1947). The objective of most of the authors to identify areas of flexibility — to discover strategies for affecting social inequality — arises from, in Gramsci's phrase, a certain "optimism of the will." It represents a view of scholarship as engaged. Objectivity in this approach becomes defined dialectically, based in a critical stance toward existing social arrangements that reinforce domination.

Given the broad-ranging character of the essays collected here, we have grouped them more for the sake of convenience than because we believe they address sharply different topics. The first piece, by

Tomaskovic-Devey and Miller, argues explicitly, as do many of the
other chapters implicitly, that national economic policy, rather than
policies overtly directed at urban development, constitutes the real
urban policy. In the United States, the consequence of that economic
policy, which they term the recapitalization of capital, will be an
increasing flow of funds to areas based in high-technology and a
continued decline of old industrial cities.

The chapters by Gough and by Adams and Freeman examine
national social policy under the two major capitalist regimes pledged
to reverse the growth of the welfare state. Gough focuses on the
overall impact of state social welfare expenditures in England; Adams
and Freeman look more closely at the effect of retrenchment in social
services in both England and the United States. Both chapters conclude that the real test of retrenchment is not economic but political;
not whether it succeeds in diverting money from "unproductive"
social expenditures into economic expansion, but whether those adversely affected will mobilize behind an alternative social strategy,
thereby destabilizing these conservative regimes.

Marcuse, while concentrating on housing policy, roots this facet of
social policy in overall responses of the German and American governments to economic crisis and popular unrest. Housing policy,
rather than responding to particular conditions in housing markets, is
devised to deal with general questions of accumulation and legitimacy. His analysis divides policy in both countries into stages, which
he explains in terms of the changing balance of national political
forces.

Third World urbanization and regional development is the topic of
the two chapters by Walton and El-Shakhs. Walton, while accepting
the overwhelming influence of the world economy on peripheral
urbanization, nevertheless criticizes dependency theory for its over-
determinism. He emphasizes the significance of national social structure, policy, and culture in differentiating outcomes among nations.
Similarly, El-Shakhs shows that while cumulating regional inequality
is a common trend of less developed countries, it is nevertheless
susceptible to governmental intervention. Again the theme is struck
that the intervention that most profoundly affects spatial development is not explicitly spatial in its intent, but responds to other policy
objectives like aggregate growth and national integration.

Our own chapter and the one by Rich both look at the response of
American city governments to economic change and political
pressure. Our emphasis is on physical development; Rich examines
public service delivery. We are particularly concerned with the politi-

cal and social conditions which permit privately financed urban redevelopment at the expense of the poor and working-class, as well as
with the significance of the increasing resemblance between "converted" American cities and their European counterparts. Rich
explores the relationship between the ways in which public services
are financed and their beneficiaries; he places local service delivery —
treated by most scholars as if each service area were financially and
functionally autonomous — in the broad context of economic power
and political influence.

Whereas the bulk of the essays start with public policy and its
outcomes, the last section takes as its starting point popular movements and then traces their effects on governmental intervention.
Perhaps because they begin with social base rather than social function, they place greater stress on noneconomic variables. Katznelson,
Gille, and Weir emphasize the significance of race rather than class in
school movements within American cities. They show the marked
differences between the United States and more class-conscious
societies such as England and argue that these differences are fundamental in their logic and outcome despite common capitalist political
economies. Castells and Murphy describe a political movement
which is simultaneously a sexual orientation and a cultural life style,
but not a class-based tendency. Rather, the development of gay political power in San Francisco depended on the capture of contiguous
territory.

Ceccarelli extends the analysis of urban political movements to a
general comparison of their development throughout Western
Europe. But although he stresses certain common stimuli that have
produced movement responses, his case study of Italy demonstrates
the substantial difference in content and role as a consequence of
differing national contexts. In particular, the links between movement demands and the programs of leftist political parties and municipal governments that exist in Italy meant that moderate sections of
these movements could be integrated into the established political
structure.

Finally, Gendrot's comparison of urban movements in the United
States and France again demonstrates that the character of national
government and traditional social values color the nature of response
to movement demands. This is the case even though the (often
heterogeneous) social composition and policy objectives of movements in the two countries may be strikingly similar.

The contributors thus place the development of urban policy
within the framework of structural analysis. But while they give

overarching importance to economic forces in determining the conditions within which governments operate, their descriptions of social structure extend beyond class analysis to a specification of national, cultural, and ideological determinants.

NOTE

1. The relationship between the state and capital has an illusory quality which is sustained by the institutional separation of "government" from "the economy." As viewed by business, the state is formally autonomous and sovereign. Capital, however, is the source of value and the market is the only rational (i.e., efficient) allocator of capital. These illusions can be so well sustained because they capture an underlying reality. Capitalist political domination allows business to determine investment and production. With the state relegated to the role of facilitating the market, mitigating its consequences, and occasionally saving some bankrupt capitalists, the state inevitably appears as necessarily unproductive. Finally, and full circle, in the name of productivity and efficiency, the capitalist state is restrained from upsetting the so-called natural functions of the market, i.e., the social relations of production which guarantee capitalist hegemony *through* the separation of economic and political spheres.

REFERENCES

ANDERSON, P. (1979) Considerations on Western Marxism. London: Verso.

CASTELLS, M. (1977) The Urban Question. Cambridge, MA: MIT Press.

FAINSTEIN, N. I. and S. S. FAINSTEIN (1979) "New debates in urban planning: the impact of marxist theory within the United States." International Journal of Urban and Regional Research 3 (September): 381-403.

GERTH, H. H. and C. W. MILLS (1964) From Max Weber. New York: Oxford University Press.

HARLOE, M. (1977) Captive Cities. London: John Wiley.

MARX, K. (1974) "Theses on Feuerbach," pp. 197-99 in K. Marx and F. Engels, The German Ideology. New York: International Publishers.

O'CONNOR, J. (1973) The Fiscal Crisis of the State. New York: St. Martin's.

Part II

**National Policy and
Urban Development:**

Europe and the United States

2

Recapitalization:
The Basic U.S. Urban Policy of the 1980s

DONALD TOMASKOVIC-DEVEY

S.M. MILLER

□ AS THE EIGHTIES BEGIN, liberalism of the Great Society/ Keynesian variety is attacked as worn out; a highly touted supply-side economics is offered in its place. A deep ideological movement toward decreasing the role of government and increasing that of business is currently underway. While the change appears sudden, it has been developing for some time and is not limited to the United States. Regardless of their merits, supply-side economics and Reaganism's opposition to an active state should not be dismissed as merely media hyperbole. They are a genuine attempt to disrupt the postwar status quo based on New Deal/Great Society liberalism and right-wing social democracy.

THE RECAPITALIZATION TURN

Recapitalization is not a formal policy that directly specifies the shifts in federal spending and taxation. It is an ideological reformulation of how and why the federal government should regulate the division of national income. In several countries, but most clearly in the United States and the United Kingdom, big business and banking interests, supported by influential economists and policy analysts (with electoral support from inflation-weary middle-income groups), have been laying out the terms of that reformulation.[1]

Recapitalization has six major objectives within the general aim of making the United States more competitive in world markets:

(1) Reduce taxation on corporations and the well-to-do in order to promote investment and thereby productivity.

(2) Contract the public social sector in order to offset the decline in tax revenues, decrease federal deficits, and reduce inflationary pressures.

(3) Increase the role of manufacturing, especially of exported goods, within the private sector.

(4) Reduce inflationary pressures by restraining expansion of the domestic economy and dampening wage increases.

(5) Decrease governmental regulation of business and industry; reduce antitrust action, especially against export-oriented firms; contract environmental, occupational safety, and consumer protections.

(6) Lessen macro intervention in the economy and rely more on regulating the supply of money.

The way to the twenty-first century is through the eighteenth. Adam Smith's 1776 guide, *The Wealth of Nations,* is again leading the way to a new era — that of restoration of an unregulated capitalism that never was. The slogans of reindustrialization and supply-side economics do not wholly convey what is being proposed. A more appropriate term is "the recapitalization of capitalism." We use "recapitalization" in a physical, economic, and ideological sense. One effort is to increase the amount of physical capital or investment in industry. A second intention is to expand the role of private economic capitalism in American society by reducing the domestic activities of American government through lowering social expenditures and by contracting the regulation of industries. The third ideological dimension of recapitalization entails the denial of obligatory levels of government support for vulnerable populations. Less government means more capitalist activity.

Recapitalization is premised on the continuing expansion of international trade and the importance of expanding exports, especially in the manufacturing sector, and on reducing imports by underselling goods manufactured in other countries. The hope is to restore American (and British) manufacturing to a more competitive place in the world economy through increased investment and reduced labor costs. An export-oriented strategy requires the strengthening of manufacturing and allied activities since few services can be exported. It is this setting of expanded international trade which shapes the basic elements of recapitalization and makes it more than the simple, long-expressed, conservative argument for untrammeled private enterprise.

The success of recapitalization depends upon the elixir of productivity. The assumption is that heightened productivity will make the United States the unparalleled leader of the capitalist world. U.S. exports will once again dominate world markets and U.S. plants will be revitalized by the infusions of productivity-enhancing high-technology; exports will grow mightily while foreign imports will wither. The trick is to get U.S. capitalists to invest in U.S. factories. Tax cuts, the easing of regulatory restrictions on business, and a contracting public sector (to make room for the expanding private sector) are inducements to corporations to produce more at lower costs.

Recapitalization entails two linked sets of processes to save the United States from the current "crisis" which is little understood but much feared. The first is an ideological (and increasingly actual) return of the economy to the guidance of the free hand of the market. The second is the redistribution of income to those people who know best what to do with it — the rich and their corporations. The goal of both is to strengthen the position of the U.S. economy in an environment of heightened international competition.

RECAPITALIZATION AND THE URBAN FISCAL CRISIS

The urban fiscal crisis (see Markusen, 1978; Tabb, 1978; Hill, 1978; Fainstein and Fainstein, this volume) was the precurser of recapitalization. In the late seventies New York City served as the Hiroshima of the welfare state, demonstrating the bankruptcy threat that faced a city profligate enough to provide services to the poor and to pay decent salaries. In both cases, the urban fiscal crisis, national recapitalization, and the social and economic contradictions produced by structural changes in economic activity led to political attacks on urban governments (e.g., NYC) and on the federal government as incompetent, bloated, and a drag on the private sector. High taxes to finance state payrolls and social services, not the movement of private capital, were identified by business, the media, and some academics as the cause of dangerous fiscal gaps in the budgetary process.

Tabb (1978) calls this diagnosis of urban problems an example of "blaming the victim" in which the classes most harmed by the actions of the market are blamed for the cost of taking care of them. Just as the fiscal crisis deligitimated the claims of the poor, who are seen as standing in the path of central city development, recapitalization is an attack on public sector activity, especially social services, which do not seem to foster private accumulation.

The urban fiscal crisis and characteristic response of fiscal austerity and investment in revitalizing downtowns have been analyzed in terms of local class politics (Fainstein and Fainstein), suburbanization (Hill, 1978; Mollenkopf, 1978; Markusen, 1978), and city-specific mismanagement (Congressional Budget Office, 1978). These explanations neglect the broad structural shifts within U.S. capitalism that have led (albeit indirectly) to urban fiscal crisis and the more ephemeral "crisis of capitalism" that recapitalization ideologies and policies address.[2]

It is true that suburbanization, the movement of upper-and middle-class residents and manufacturing plants from the central city to the suburbs, undercut the city tax base which supported expenditures on urban infrastructure and social services. But this loss/burden approach is only a partial explanation of the budgetary gap that characterized urban fiscal crises. The fiscal crises of older cities are only one result of changes in the structure of economic activity. Women entering the labor force; the rise of the sunbelt and decline of the frostbelt; the rise of service industries and the relative decline of manufacturing; and the internationalization of the U.S. market produced a macro-economic shift in the employment, product, and regional composition of the U.S. economy.

STRUCTURAL CHANGE AND OLDER URBAN AREAS

Three different but related processes of national economic change disturbed the older urban areas of the Northeast and Midwest and contributed to fiscal gaps and new social problems.

Deindustrialization. There is a pattern within capitalism of accumulation and disaccumulation which is probably best-known in the fluctuations of the business cycle, but which is also apparent in the long-wave Kontradief cycle (Castells, 1980) — what Galbraith (1980) has called the "aging process." Older urban centers became relatively less profitable in the late 60s and 70s (due to older factories, higher wages, higher taxes, and higher energy costs) resulting in plant shutdowns and disinvestment (Bluestone and Harris, 1980). Capital flowed elsewhere — first to the suburbs, but more recently to the U.S. Southwest and overseas. Eventually, when property values, taxes, wage rates, and so on become sufficiently depressed, capital investment may return to the Northeast. Fainstein and Fainstein (this volume) point out that some central cities are again becoming profitable places for investment. The process is evident in New Hampshire and Massachusetts, where more manufacturing jobs were created than lost in the late seventies due to, among other factors, their relatively low (for the Northeast) wage rates and various local tax incentives. There is no guarantee, however, that depressed areas will soon become prosperous again or that the content of any new prosperity will favor the same classes or industries as are presently declining.

Uneven Development. Major capitalist firms are not leaving urban centers so much as growing elsewhere and imposing a new division of labor. The decline of the older industrial regions of the United States is relative to the rapid growth of the new areas; only occasionally is the decline absolute. The process is one of growth and geographical dispersion as new products are developed, new labor forces exploited, and new markets invaded. With geographical dispersion comes a geographical division of labor. The United States as a whole and the world in general are becoming regionally specific for production (and even consumption). The older urban centers are becoming *control* centers (e.g., corporate headquarters, banking, law firms) for production processes that are often located elsewhere and *consumption* centers for the corporate elite. This makes for great inequality within the older urban centers as middle-range jobs locate outside the cities, leaving low-paid service sector employment and high-paid corporate control sector employment. The uneven development of U.S. capitalism deprives the older urban centers of an adequate supply of good jobs and widens inequalities.

Service Society. Within metropolitan areas and in the U.S. economy in general, investment and employment are shifting from manufacturing into services, as well as from urban centers to southern and foreign locations. The American product line is changing. Services to manufacturing are growing in the form of real estate, banking, and law and accounting firms as are crucial middle sectors between manufacturing and consumers such as transportation and wholesale trade. Services to consumers are growing in the form of retail trade, restaurants, entertainment, and so on. Comments on these changes abound; understanding lags. Some commentators (Bell, 1973) have characterized these industries as providing high-knowledge employment, but the reality is that most service jobs are low-paid, heavily populated by women and minorities, and nonunionized. Dependence on these new products — services, information, control — is precarious because they can be relocated easily due to their low capital investment (at least compared to manufacturing). On the other hand, they do and have produced the majority of new jobs in the U.S. economy over the last decade.

These economic shifts have had the largest and most deleterious effects upon the working classes of the older urban industrial belt. The urban fiscal crisis is only one result of these economic shifts that threaten the cities of the Northeast and Midwest. Others include: the permanent loss of high quality employment when manufacturing jobs are not replaced or are replaced by low-paid, low-quality service jobs; rising inequalities between a small high-paid corporate elite and a large pool of marginal service workers; a rise in the proportion of older residents as younger people move south and west following the jobs; a general decline in living standards, services, and morale as economic prospects are mired at new low levels. The structural shifts in the U.S. economy are producing a host of present and potential social problems for the older industrial cities.

ECONOMIC POLICY AS THE BASIC SOCIAL POLICY

National economic policy shapes all other governmental attempts to influence the social and economic welfare of citizens. Thus, economic policies that promote fast economic growth provide the working classes with increased bargaining power over employers. Corporate tax depreciation schedules that favor new factories over old help the southwestern United States but hurt the Northeast, as new factories are built in low wage areas. High interest rates set by the Federal Reserve cause construction industry depression while allowing bank profits to soar. The distribution of tax burdens determines who loses income in order to pay for government programs including transfers, social and physical infrastructure, and future economic growth.

Federal social objectives are routinely subordinated to federal economic policy. For example, local urban planning depends upon economic resources, both from federal sources and local taxation. As fewer and fewer resources are available to local governments in urban areas, planning and policy are limited to managing the most necessary urban services — police, fire, schools, mass transport — while less necessary services are curtailed. At this point, a point already arrived at by some cities such as New York and Boston, most other services are left to the private sector, to the hidden hand of the market. Jobs, housing, training, and the distribution of income beyond the (shrinking) government baseline will, at least in the present context of recapitalization, increasingly be left to the market to provide. The role of urban planner will be assumed by federal economic policy, particularly policies of taxation, spending, and interest rates, in conjunction with market forces (i.e., the decisions made in the private sector).

Economic policies tend to be prior to social policies just as economic relations tend to be prior to social relations in capitalist economy. In some senses, then, transfer programs, urban planning, and social goals within state institutions are constrained by the availability of funds and the smooth functioning of the capitalist economy.[3] The great gains in social programs and revenue-sharing in the sixties, although encouraged by political activism, were possible because of the relatively fast growth of the economy and federal receipts. Social programs like Aid to Families with Dependent Children (AFDC) and Manpower (later Employment and Training) programs were the product not only of valuable social goals but also of a political willingness and economic flexibility to make room for social transfers in the federal budget. Economic policy on the federal as well as local level is not a necessitarian or technocratic response to objective economic situations. Economic policy like social policy is the result of political forces and decisions that favor one group over another. Now, as federal spending and taxation have become politically unpopular and more of the national income is being allocated to business and the rich, social programs shrink within the federal budget.

Our general theoretical point is that economic policy is always the basic social policy. In times of shrinking budgets and fiscal restraint, this becomes much more apparent, as it is presently in the United States and in Great Britain. The role of urban planning must fit within the fiscal limitations of economic policy. Urban planning becomes more and more difficult as funds for all but the most crucial city services dry up. When this happens economic policy is not only the basic social policy but increasingly the only social policy. The distribution of employment, housing, and income are then increasingly determined by the private market place as mayors and urban planners are stripped of their budgets and isolated from their political constituencies. In the present U.S. context, the recapitalization of U.S. capitalism has become the goal of federal economic policy. Recapitalization policies will come in many cases to substitute for more conscious urban planning.

RECAPITALIZATION AS SOCIAL POLICY

The broad outlines of recapitalization policies are being enacted in the United States and the United Kingdom. Less government spending relative to overall economic activity, especially on social transfers, was embodied in the 1982 Reagan budget (as it has been in the Thatcher budgets). Local initiatives such as Proposition 13 in California and Proposition 2½ in Massachusetts have deferred spending at subnational levels. Tax cuts for corporations and the wealthy are the mainspring of the Reagan productivity-growth program. Monetarism, with attendant high interest rates and slow growth in the economy, is the inflation-fighting policy of choice. Although at this writing, no policies to dampen wage increases in the private sector have been put forward, such efforts are likely in the near future, justified in the name of export competitiveness and inflation-fighting.[4]

What effect will these policies have upon U.S. workers, particularly those in the older industrial regions of the Northeast and Midwest who bear the brunt of the strain fostered by shifts in the structure of the economy? How will recapitalization-inspired policies act as social planning for U.S. workers and the frostbelt? The answer to this question depends upon the ultimate nature and course of recapitalization as an ideological movement. The present recapitalization-inspired policies may in fact be short-lived, replaced in a new political moment by other policies. On the other hand, they may endure and constitute a new status quo.

The recapitalization ideology leads to two basic policy alternatives. The first, which has been strongly embodied in the Reagan administration's economic and social policies, is a simple upward redistribution of national income through tax "reforms" with little or no government control over what is done with the increased income of corporations and the wealthy. The market and its subtle mechanism for distributing wealth and opportunity is reembraced as the path to a stronger U.S. position in the world economy. In some sense, this is a deinstitutionalization of the class struggle over the distribution of national income. When tax breaks substitute for rising wages, the struggle over real income becomes masked as the source of income moves from employer to the government.

The second alternative may flow from the first when the hidden hand of the market proves disappointing. The upward redistribution of income will not be an effective response to structural shifts in the economy. Eventually the focus of economic policy may shift to a more targeted approach to increase productivity in the manufacturing sector, bolstering U.S. exports, and limiting U.S. imports. In both approaches, the role of government in the economy will be reduced and the share of the national income going to business will be increased. In the first alternative, the disposition of that capital is left up to the corporations and the wealthy; in the second, some form of targeting channels investment into productivity and export-related fields. The economy's response to current recapitalization policies will determine whether they remain in their present untargeted form or move into a more goal-oriented form as untargeted policies fail to fortify the U.S. economy in world-level competition.

A third possibility is that recapitalization as an economic ideology may lose legitimacy by political resistance (a precursor of which can be seen in the violent and sustained protest against similar policies in Great Britain). Resistance could come from the working classes who, at least in the short run, stand to lose, or from sectors of business that may resist targeting because it bypasses them (e.g., oil) or who might suffer from recapitalization policies in general (e.g., small businesses).

We have discussed elsewhere the inadequacies of the current recapitalization policies, who will benefit and who will lose, and the likelihood of political resistance (Miller and Tomaskovic-Devey, 1981a; 1981b). For the remainder of this article, we will discuss the consequences of recapitalization for United States workers and for older industrial regions.

REAGAN-STYLE REDISTRIBUTION

The recapitalization policies, begun in the Carter administration but vastly accelerated under Reagan, consist mainly of an upward redistribution of national income. They expand inequalities and threaten the poor, the old, and the near-old, as well as working people who find their bargaining position weakened in an atmosphere of economic crisis orchestrated to legitimate draconian policies. Even before Reagan, real incomes for the average worker and people dependent upon government transfers (including social security-dependent retirees after 1979) had been declining, while business profits, although volatile across industries, had been slowly rising (Miller and Tomaskovic-Devey, 1981b). The long-term effect of these policies depends upon what corporations and the wealthy do with their capital windfall. The expressed Reagan expectation is that they will invest in productive ventures of one sort or another, preferably in manufacturing, and boost U.S. productivity and international competitiveness.

The more likely result is that the freed funds produced by untargeted economic policies will be invested and consumed in patterns similar to those of the recent past. Many corporations will use their firm's additional funds to buy other corporations rather than risk building new plants in an environment of slow growth and uncertain demand. Capital will continue to flow into speculation of one sort or another, including real estate, commodities, and currency markets. New investment will continue to be concentrated in those industries with the highest profit potential. Investment has flowed increasingly into service industries where demand has in the last few years grown rapidly, even in times of national recession.[5] Within manufacturing, investment has favored profitable industries such as energy or microelectronics over less profitable but more needy industries such as steel and automobiles.

For workers (and consumers) this means continued inflation of housing costs, development of few jobs (if any) in manufacturing, and expansion of jobs in the low paid service sector. Capital investment in manufacturing (if and when it occurs) will take place in low-wage low-tax areas such as the Southwest and overseas, draining jobs and capital from older regions. The older cities of the Northeast and Midwest, areas of urban poverty and aging factories, will be characterized by a general stagnation of jobs and tax base, perhaps accompanied by more of the urban renewal-led expansion of corporate control functions downtown, driving up real estate prices and providing a few good jobs (e.g., lawyers, accountants, and executives), leading to local gentrification. The service sector, particularly services for the downtown business elites, may grow, providing mostly low-paid, dead-end employment. Recapitalization policies will encourage states and cities to compete for investment and jobs. Urban enterprise zones, already proposed by the state of Connecticut, and favored by Reagan et al., may be the form this competition takes in the older industrial cities and states. Urban residents will experience a further erosion of their standard of living due to cutbacks in city services to residents as budgets are pared to compete with low-service, low-tax cities in the South and West.

The present economic policies of upward redistribution and market freedom are inherently unstable. They do not address the structural shifts in the economy, including the increasing importance of exports, the decline in manufacturing and rise in services, and the fiscal crisis of older urban centers. This tax-based bribe of the wealthy will probably not endure long as a political stabilizer, because it burdens many without producing compensating benefits such as a rise in manufacturing exports (and employment) or economic stability. It is possible that the present form of recapitalization may work, but only if capitalists start acting like capitalists rather than speculators; investing in potentially productive businesses rather than attempting to make money out of money. More likely, present recapitalization policies will fail and be replaced by more targeted policies.

TARGETING ON PRODUCTIVITY

Productivity is an overburdened panacea charged with expanding exports, profits, and employment while defeating the scourge of inflation. If present redistribution policies fail to produce a supply-side-inspired burst in productivity, and consequently exports, political attention will be directed at those industries identified as the wave of the future. More attention will be placed upon "high-technology" industries such as computers, microelectronics, and precision instruments. The products of these industries are doubly attractive because of their high export potential and their usefulness in raising U.S. productivity in other industries (e.g., robotics on automobile assembly lines and word processors in the office).

If we take a best-case scenario in which the politics of recapitalization are actually transformed into economic policy targeted on productivity-enhancing high-technology, what will that mean in terms of employment and regional growth?

DEFINING HIGH TECHNOLOGY

The phrase "high-technology" is not only a categorization of industries or technologies; it is also political advertising for U.S. industry. High-technology as a gleaming phrase is not dissimilar to the familiar advertising tags of "New" or "New-improved." As technologies become suspect in an increasingly environmentally conscious age, and as traditional manufacturing industries decline, U.S. technology (and industry) need relegitimating. Suddenly, political rhetoric and economic analysis point to two kinds of technology: plain, old, dirty *technology,* and new, improved, clean *high technology.* The celebration of high technology as the wellspring of increased productivity means that industries lucky enough to be so labeled will be the main beneficiaries of recapitalization policies targeted at productivity and exports.

The search for the inner logic of high technology is circuitous and frustrating. Experts define high-technology industries in different ways. A simple criterion for high-technology industries was proposed by Barnaby Feder (1981), science writer for the New York *Times*. He characterized as high-technology those fields in which technological change is occurring rapidly. Unfortunately, he did not provide a list of industries that fit this definition or explain how to measure rapid change. In any case, the list would not be very useful over time as new industries slowed down and old ones sped up in introducing technology.

Another more measurable definition of high-technology industries uses a criterion of professional employment. The labor force of high-technology industries is characterized by an above-average percentage of scientists, engineers, and technicians (Vinson and Harrington, 1979). This definition (sci-tech definition) requires that scientists, engineers, and technicians make up 13.7 percent or more of total industry employment. This definition covers 15 industries (classified by three-digit standard industrial classification [SIC] codes of the U.S Commerce department).

Various industry and academic witnesses at a Congressional hearing (U.S. Congress, 1978) presented a number of undefined lists of high-technology industries. Although industries on the various high-technology lists overlapped substantially, many differences appeared. One definition with a corresponding list of qualifying industries was given at the hearings (U.S. Congress, 1978: 295). This definition suggested that high-technology industries could be identified by their above average research and development (R&D) intensity. These industries were isolated by dividing R&D by value of shipments, yielding a list of 19 industries that spend a lot of money on R&D and presumably are high-technology (R&D definition).

A list of 20 (again 3-digit SIC) high-technology industries was developed by economists working for the Commonwealth of Massachusetts (MA definition) which is betting its employment future on the growth of high-technology industries. Although no formal criteria are set out for the list, the authors (Massachusetts Division of Employment Security, 1981) indicate that "in general, the companies classified in these industries are labor-intensive, with a highly skilled employee base. In particular, they have a high ratio of scientists and engineers to their total labor force."

These three lists of high-technology industries encompass twenty-five industries; only eight meet all three definitions. The nebulous quality of the high-technology concept makes the choice of various definitions political decisions (see Table 2.1).

We will look at the recent growth of high-technology industries as a means for predicting their future growth. Table 2.2 compares the number of jobs added between 1969 and 1979 in high-technology industries, as variously defined, with the additional employment in the manufacturing sector, the private sector, and the overall economy. It is striking how few of the total number of new jobs in the seventies were created by the three sets of high-technology industries. Job generation ranges from a low of 71.6 thousand for the sci-tech definition with a growth of employment of only 2 percent to a high of 169.1 thousand jobs with a growth of 5.1 percent for the MA definition.

TABLE 2.1 Twenty-Five High-Technology Industries Classified by Three Definitions

Def.	SIC	Name
@#	281	Industrial Chemicals
@#	282	Plastics
@#$	283	Drugs
#	287	Agricultural Chemicals
$	348(19)	Ordnance[a]
@#	351	Engines and Turbines
@#$	357	Office and Computing Machines
@ $	361	Electrical Distribution Equipment
@ $	362	Electrical Industrial Apparatus
$	363	Household Appliances
$	364	Electric Lighting & Wiring Equipment
#$	365	Radio & T.V. Equipment
@#$	366	Communications Equipment
@#$	367	Electronic Components
$	369	Misc. Electrical Equipment, Supplies
@#	372	Aircraft and Parts
@ $	376	Space Vehicles & Guided Missiles[b]
$	379	Misc. Transport Equipment
@#$	381	Engineering and Scientific Instruments
@#$	382	Measuring and Control Instruments
@#$	383	Optical Instruments
#$	384	Medical Instruments
#$	385	Opthamolic Goods
@#$	386	Photographic Equipment and Supplies
#$	387	Watches, Clocks

@ MA definition.
# R&D definition.
$ Sci-Tech definition.
(For more on definitions see Table 2.2 and text.)

[a]The SIC code for ordnance was changed from 19 to 348 in 1972.
[b]Data not available prior to 1977.

Compared to the 19 million jobs and growth of 27.6 percent in employment in the economy as a whole, high-technology does not seem like a promising job generator. Taken together, the three definitions suggest that in terms of employment, the high-technology industries produced jobs at about the same slow rate as manufacturing in general. The process of job generation by high-technology firms is, however, more complex, as is indicated by the inclusive (24 industries) and exclusive (9 industries) definitions of high-technology industries that can be formed as a composite of the 24 industries in Table 2.1.

TABLE 2.2   Employment Levels in High Technology Industries (various definitions), Manufacturing and the Private Sector, U.S. 1969-1979

	(in thousands)			1969-1979 absolute change	percentage change 1969-1979	percentage change 1974-1979
	1979	1974	1969			
9-industries[a]	2132.8	1623.2	1688.8	443.9	26.3	28.6
R&D def.	3622.8	3269.3	3467.9	154.9	4.5	10.2
Sci-Tech def.	3645.9	3288.5	3574.3	71.6	2.0	9.5
Mass. def.	3492.0	3325.4	3322.8	169.2	5.1	5.0
24-industries	4694.1	4596.0	4834.0	−139.9	−2.9	2.1
All manufacturing	20972	20016	20121	851	4.2	4.8
All private	73870	64050	57914	15956	27.6	15.3
All employment	89482	78334	70141	19341	27.6	14.2

SOURCE: U.S. Bureau of Labor Statistics, Employment and Earnings
   Vol. 27, 3 (March 1980) Table B-2, pp. 58-70;
   Vol. 16, 9 (March 1970) Table B-2, pp. 44-58;
   Vol. 21, 9 (March 1975) Table B-2, pp. 52-64.

[a]The three definitions of high technology include a total of 25 industries; for 1 of them (SIC 376), data are not available prior to 1977. Thus, "24-industries" is the largest sample of high technology industries; "9-industries" is made up of the 8 industries that overlapped on all 3 definitions plus SIC 385 which was combined with SIC 383 in the 1969 and 1974 data sets. The R&D definition employs a criterion of R&D as percentage of value of shipments; the sci-tech definition employs a criterion of scientists, engineers, and technicians as percentage of employees; and the MA list actually has no formal definition, but was selected by a group of economists-experts in the Massachusetts Department of Employment Security. See text for more information.

If we accept as high-technology all 24 industries included in any of the 3 definitions, a *net loss* of about 140 thousand jobs occurred in this sector in the seventies. A tremendous drop in employment in a few industries produced this contraction. Industrial Chemicals (SIC 281) declined by 142.9 thousand jobs, ordnance (SIC 19 in 1969 and 348 in 1979) decreased by 267 thousand jobs, and aircraft and parts (SIC 372) lost 200 thousand jobs over the 10-year period. In addition, plastics (282), electrical distribution equipment (361), household appliances (363), radio and TV (365), engineering and scientific instruments (363), and watches and clocks (387) all reduced employment over that period. The employment gains in the other 15 industries did not offset the losses in these 9. Explanations for this net loss include the decline in war purchases by the U.S. government after the Vietnam war for SICs 348 and 372. For the others, some combination of loss of demand to foreign competition (e.g., SIC 365) and productivity-

TABLE 2.3 Growth in Value Added and Employment in U.S. High Technology Industries (various definitions) and Manufacturing Sector.

	1967-1977 Percentage Growth Value Added	1969-1979 Percentage Growth Employment[a]
9-industries[b]	144	26
R&D def.	97	5
Sci-Tech def.	132	2
Mass. def.	88	5
24-industries	99	−3
All Manufacturing	123	4

SOURCE: Computed from *Statistical Abstract of the United States; 1980*, Table 1451, pp. 814-819 and Table 2.2 of this article.

[a]Value added data is not presently available for all industries in the various high technology definitions after 1977. The 1967-1977 and 1969-1979 data for value added and employment should be treated as a rough comparison.
[b]See Table 2.2 for explanation of various high technology definitions.

related displacement of workers (e.g., SIC 281) led to a decline in employment.

Defining high-technology industries as only those industries covered by all three definitions suggests a much rosier employment potential. These 9 industries grew by 443.9 thousand jobs. Although this is an insignificant proportion of total employment growth (2 percent over the 10-year period), the growth rate was similar to that of the economy in general and much higher than that for manufacturing as a whole. This finding should be understood in terms of the *variability* of manufacturing employment. As some sectors of manufacturing add employees, others contract. This pattern holds for the glamor industries that have recently been called high-technology as well as for manufacturing overall.

Much of the small growth in high-technology employment has been concentrated in the most recent years (post-1974). There has been some acceleration in job creation by these firms in the late seventies. High-technology employment has been growing in recent years, but it is only now catching up with the high levels of the Vietnam period.

The record of high-technology firms in generating jobs is not reassuring. The growth in demand for high-technology goods has far outstripped the growth of high-technology employment. In the 1970s, the output of high-technology goods (value added), using the most restricted and favorable definition (9 industries), grew by 144 percent, while employment grew by only 26 percent over the same period. It takes a tremendous increase in high-technology output to produce a small increase in employment.

During the seventies, the value added for all manufacturing increased by 123 percent, a substantially slower increase than in the growth in demand for the most favorable definition of high-technology industries. If we use the MA definition of high technology, however, for comparison with manufacturing demand and employment, then high-tech growth did not exceed the stagnant manufacturing sector of the economy. Overall, manufacturing value added grew by 123 percent and employment by 4 percent; high-technology industries, using the MA definition, had an expanded product of only 88 percent and an employment increase of 5 percent.

It takes a tremendous growth in demand to make any increase in employment for high-technology industries. Depending upon the definition of high technology, there have been variable increases in demand over the decade. The prospects for job creation are dimmed further by the possibilities of job loss in other industries that use the productivity-enhancing tools produced by high-technology firms.

As Table 2.2 shows, of the more than 19 million jobs added to the U.S. economy between 1969 and 1973 by the fastest growing definition of high-technology (the 9 industry version), less than 3 percent (440 thousand jobs) was attributable to employment expansion in high-technology fields. Even going from the 1974 trough in high-technology employment to 1979, it still furnished less than 5 percent (510 thousand jobs) of the growth of total employment.

The limitations of high-technology employment are not only quantitative, they appear also in its composition. High-technology firms employ proportionally fewer production workers and minorities than do manufacturing and the private sector (see Table 2.4). These are, however, the groups which suffer high unemployment. Growth in high-technology employment, if it does occur, will not necessarily help those groups that need help the most.

Table 2.5, on the regional distribution of high-technology employment (again using the most restricted definition, now down to seven industries: see footnote Table 2.5), shows that the older industrial areas with their high unemployment rates actually lost high-technology jobs between 1972 and 1976 (the latest available state level data). There are two things immediately apparent from the table. The first is that much high-technology employment is in the older industrial states. The second is that even as high-technology employment

TABLE 2.4 Production Workers, Women, and Minorities as a Percentage of High Technology Industries (9-industry definition), Manufacturing Sector, and Total Employment.

	Percentage[a] Production Workers	Percentage[b] Women	Percentage[b] Minorities
High-Tech	50	36	8
Manufacturing Sector	72	31	11
Total Employment	82	41	11

SOURCE: U.S. Bureau of Labor Statistics, *Employment and Earnings*, March 1980, Vol. 27-3, Table B-2, pp. 58-66.
U.S. Bureau of Labor Statistics, *Employment and Earnings*, January 1979, Vol. 26-1, Table 30, pp. 179-180.
[a]Data for 1979.
[b]Data for 1978.

expanded across the entire United States, there was a net loss in the older industrial states. There is little hope, given the pattern of regional employment growth between 1972 and 1977, that significant high-technology-generated employment benefits will occur in the older industrial cities.

Competition with the union-free, low-wage, low-tax South and West will be difficult for older industrial areas with their expensive urban infrastructure, high wages, and large poor populations. The New England region is a useful example, however, of how a once-declining region can compete for manufacturing jobs. Massachusetts, in particular, has endeavored to improve the "business climate" for high-technology firms by limiting property and business taxes as well as spending a good deal of money on advertising trying to attract firms. Massachusetts was, however, in a favorable position to attract these firms because it already had a reputation as a high-technology state, as well as a highly skilled labor force and an attractive physical environment.

Clearly, then, high-technology industries have not produced a reassuring number of jobs. Moreover, the use of high-technology products in other industries may eliminate as many or more jobs than are created by their production. Those jobs that do result from targeted federal policies will not necessarily be in those regions where jobs are most needed or for those workers who have the hardest time finding useful work.

The hope that tax cuts will enhance savings, investment, productivity, sales, and employment rests to a major extent on the expan-

TABLE 2.5   High Technology Employment (7-industry definition) by
            Region, 1972-1977.[a]

Region[b]	(in thousands) 1977	(in thousands) 1972	Absolute change	Percentage change
New England	166.3	136.9	30.6	21.4
Older Industrial	447.1	459.7	−12.7	−2.8
South	99.3	78.4	20.9	26.6
West	351.6	261.6	90.0	34.4
U.S.[c]	1590	1396	194	13.9

SOURCE: U.S. Census of Manufactures, 1977.

[a]This definition of high technology included SIC codes 283, 357, 366, 367, 382, 383, 386.
Code 381 was omitted because employment was so low in surveyed states that it was not
included in the census's regional tables.

[b]A sample of states was taken for each region. The sample included all states that had high
technology industries among their top five manufacturing industries plus a few large states
with substantial employment. New England includes Massachusetts, Connecticut, and
New Hampshire. The older industrial region includes Michigan, New Jersey, Indiana, Il-
linois, and New York. The southern region includes Georgia, W. Virginia, North Carolina,
Florida, Tennessee, and Alabama. The western region includes the states of Arkansas,
Louisiana, Texas, Arizona, Colorado, California, and Washington.

[c]Data for United States is for the whole country.

sion of the high-technology sectors of industry. Our discussion
demonstrates that this expectation is shaky, at least in its employment
possibilities. Large cities, outside of the American West, are unlikely
to benefit from recapitalization policies and the envisioned growth of
productivity-enhancing industries.

CONCLUSION

Recapitalization policies aimed at strengthening the U.S. position
in the world marketplace through the release of capital to the wealthy
and their corporations and the curtailment of state funding for social
programs will severely limit the possibility of active urban planning.
Attempts to offset the social and economic problems generated by
deindustrialization, uneven development, and the rise of the service
sector are already less and less practical as government funds are
redirected into bolstering the profitability of private enterprises.
    Within the context of limited planning possibilities, the economic
policies of the federal government and the resultant market developments in the private sector will become the de facto urban planners of

U.S. cities. This is less true for cities that are presently experiencing rapid growth and increasing tax revenues. For the older industrial cities, like New York and Detroit, recapitalization policies and market forces will largely determine employment, tax base, and effective demand for their local economies.

Declining federal funds for social programs and transfer mechanisms will limit effective demand and growth in these older cities characterized by high unemployment and large poor and old populations. It will be left up to the market to provide jobs and tax base as federal funds shrink as a source of both jobs (e.g., CETA) and small business development (e.g., CDC funds). This is certainly true in the untargeted Reagan form of recapitalization. If increased targeting on productivity and exports leads to government subsidies for high technology industries, both jobs and taxes will be generated. However,there will not be many jobs; they will not tend to be located in older industrial states; and the products of these industries may tend to eliminate jobs in other sectors of the economy as robotics and wordprocessors replace autoworkers and secretaries.

The return to free market economics and the upward redistribution of national income implied by the recapitalization of U.S. capitalism will almost certainly fail to bribe private enterprise to invest in the U.S. economy. Even if recapitalization were to succeed in producing a much stronger or secure place for the United States in international trade, its consequences as urban planner of first resort will be negative for the older industrial cities that are most affected by the current transformation of the U.S. economy.

NOTES

1. France, with its recently elected socialist government, stands alone among advanced capitalist nations with economic policies premised on rapid growth and an active welfare state.

2. We use "crisis of capitalism" hesitantly because the crisis is little more than a mismatch (a contradiction in Marxist terminology [Castells, 1980]) between existing political and social institutions and the evolving economic structure and only a "crisis" because the changes have been viewed with such alarm. Crisis then becomes at least as much a political tool as a description of economic problems. The crumbling of the capitalist world economy is not at hand, although tensions between the regions, sectors of business, and classes need to be addressed. Recapitalization is one form of response to changes in the economic structure; it is by no means the only possible response and is certainly not desirable from the standpoint of the poor and working classes.

3. Of course, social relations interpenetrate and shape economic relations as well.

4. The Reagan administration's handling of the air traffic controllers strike might be an indicator of its general intent.

5. Our explanation for increased demand of services even in recessions is threefold. The speculative activity of business increases in poor economic times and so accountants, lawyers, and other business services prosper. Service industries require less capital investment than manufacturing and are attractive investments for that reason. As individual earnings decline, more members of the family must enter the labor force and so the demand for consumer services like restaurants and dry cleaners also increases. This trend, however, may falter if consumer demand is choked off by the upward redistribution of income.

REFERENCES

BELL, D. (1973) The Coming of Postindustrial Society. New York: Basic Books.

BLUESTONE, B. and B. HARRISON (1980) Capital and Communities: The Causes and Consequences of Private Disinvestment. Washington: Progressive Alliance.

CASTELLS, M. (1980) The Economic Crisis and American Society. Princeton, NJ: Princeton University Press.

Congressional Budget Office (1978) "New York City's fiscal problems," in R. Alcaly and D. Mermelstein (eds.) The Fiscal Crisis of American Cities. New York: Vintage.

FEDER, B. (1981) "Technology: defining terms in growth area." New York Times (April 2): D2.

GAILBRAITH, J. (1981) "The conservative onslaught." New York Review (January 22): 30-36.

HILL, R. (1978) "Fiscal Collapse and Political Struggle in Decaying Central Cities in the U.S.," in W. Tabb and L. Sawers (eds.) Marxism and the Metropolis. New York: Oxford University Press.

MARKUSEN, A. (1978) "Class and urban social expenditures," in W. Tabb and L. Sawers (eds.) Marxism and the Metropolis. New York: Oxford University Press.

Massachusetts Division of Employment Security (1981) "High Technology Employment in Massachusetts and Selected States," Commonwealth of Massachusetts Labor Area Research Publications.

MILLER, S. M. and D. TOMASKOVIC-DEVEY (1981a) "A Critical Look at Reindustrialization." Social Policy (January/February).

——— (1981b) The Recapitalization of Capitalism (forthcoming).

MOLLENKOPF, J. (1978) "The postwar politics of urban development," in W. Tabb and L. Sawers (eds.) Marxism and the Metropolis. New York: Oxford University Press.

TABB, W. (1978) "The New York City Fiscal Crisis," in W. Tabb and L. Sawers (eds.) Marxism and the Metropolis. New York: Oxford University Press.

U.S. Congress, Subcommittee on International Finance of the Committee on International Finance of the Committee on Banking, Housing and Urban Affairs. (1978) Hearing on "Oversight of U.S. high technology exports." Washington, DC: U.S. Government Printing Office.

VINSON, R. and P. HARRINGTON (1979) Defining High-Technology Industries in Massachusetts. Boston: Commonwealth of Massachusetts, Department of Manpower Development.

3

The Crisis of the British Welfare State

IAN GOUGH

□ AT THE TIME OF WRITING (July 1981) the chickens are coming home to roost. Some of the most serious urban riots seen in mainland Britain (excluding Northern Ireland) this century have swept parts of London, Liverpool, and other major cities during the last month. The most significant challenge to Thatcherite policies thus far has come not from the organized labor movement, nor from within the Conservative party and the ruling-class establishment, but from the spontaneous protest of black and white youth in the deprived centers of Britain's decaying cities.

In October 1980, Professor J. K. Galbraith wrote that Britain was the ideal country in which to expose the inanities of monetarism. "Britain is the Friedmanite guinea-pig. There could be no better choice. The British do not take easily to the streets." This has now been disproved, but in a way that reveals the contradictions of Thatcherism in a clearer light. As I argued in an earlier paper (1980),[1] a policy of monetarism and antiwelfare statism exacerbates all three contradictions that the Keynesian welfare state attempted to overcome: barriers to capital accumulation, increasing needs for social reproduction, and problems in securing political legitimation. Yet there can be no going back to the economics of Keynesianism, the social policies of Fabianism, or the politics of centerism. The necessity of developing a "third way" has become even more urgent. Though it is not my purpose here, I hope this chapter can contribute toward a new approach by exposing the contradictions of both Thatcherism and its predecessor, social democratic reformism, in the British context from the perspective of Marxist political economy.

THE NEW CONSERVATISM

Since the mid-1970s, two notable shifts have taken place within the advanced capitalist world. First, the economic situation deteriorated sharply and the outlook for long-term capital accumulation and economic growth has become bleak. The OECD countries as a whole are assailed by a combination of economic problems unprecedented since World War II: 23 million out of work according to unreliable official statistics, inflation rates averaging 10%, a combined current account deficit of $75 billion, and a fall in their combined GNP in the second half of 1980 (OECD Biannual Report, quoted in The Economist, October 11, 1980). Behind these trends lies a longer-term fall in the rate of profit since the 1950s (Heap, 1980/1981).

Second, the governments of several major countries have shifted to the right, including the United States, Britain, Australia, New Zealand, Sweden, and Holland. Only West Germany, Austria, the other Scandinavian states, and now France,[2] are presently governed by social-democratic parties or coalitions. Furthermore, the conservatism espoused by many of these states is of a qualitatively different kind from that witnessed since World War II. One common feature of this "new conservatism" is a strident attack on the bundle of citizenship rights and social benefits referred to as the welfare state. It is the implications of this rightward shift for the future of the welfare state with which this chapter is concerned.

But despite many similarities, these countries exhibit considerable variation in their economies, state structures, welfare policies, and political and ideological practices. Hence, an understanding of the politics of the welfare crisis requires detailed study of each nation-state: This analysis will focus on Britain and the politics of Thatcherism. Superficially similar to the Reagan administration in the United States — in its antidétente, Cold War foreign policies, its appeal to reactionary values, its antiegalitarianism and its individualism — there yet remain significant differences between the two, especially in the field of economic policy where the Thatcher government still espouses explicit monetarist prescriptions. In my view, a study of the "Thatcher experiment" by welfare movements abroad will yield dividends in political and theoretical understanding. Politically, it serves as an awful warning of the consequences, dangers and contradictions of the policies of the new right. Theoretically, it provides an important lesson about the way political and ideological factors mediate the economic crisis and welfare crisis in different countries.

In many respects Thatcherism is a paradigmatic case of the new conservatism, and this is because the combination of problems faced by the British state is more extreme and intractable than elsewhere in the Western world. The British economy has suffered a long-term decline over many decades. The capitalist class has never been able to implement a coherent strategy to arrest this decline, due in part to a persistent split between industrial capital and finance (and now oil) capital. It is faced by a long-established, relatively unified, economically strong, and defensively oriented working-class movement. Finally, it has a well-established and in some respects egalitarian welfare state.[3] These factors, plus others, mean that the new conservatism in Britain has been forced to develop a more explicit ideology and political strategy than, for example, in the United States. Capital in the United States can move to the sunbelt states to avoid and weaken the frostbelt's trade unions and more generous welfare systems. Such internal mobility is nigh impossible in Britain: Public assistance benefits are identical in the Shetland Isles and in London, apart from housing cost allowance. And the level of unionization (over 50 percent of the workforce compared with less than 20 percent) has forced the new right to tackle explicitly trade union rights and living standards, that is, working-class power and wealth.

What then has been the record of this new political tendency in Britain? How and why has it sought to retrench and restructure the British welfare state? How can one characterize the ideology of Thatcherism? What are the new contradictions these policies are engendering, and what new conflicts may we expect to witness? Finally, what are the longer-term prospects of the new right in Britain? These are the major questions I want briefly to tackle here.

RETRENCHMENT IN THE BRITISH WELFARE STATE

The first point to make is that the current crisis in the welfare state has a quantitative and qualitative dimension. The former process — economic cutbacks in social expenditure — was begun by the Labor administration in 1975. The latter — the ideological attack on the British welfare state — was more specifically initiated by the new Conservative administration elected in May 1979.

THE CUTS

In the summer of 1975, the Wilson government moved sharply to the right, renounced the social contract with the trade union movement, and introduced a sharp cutback in government expenditure plans. Between the financial years 1975-1976 and 1977-1978, total social spending actually fell by 0.4 percent in real terms, following a purge of most capital programs in housing, health, education, and welfare, and despite a strong rise in spending on social security (see Table 3.1). A new system of "cash limits" for the first time enabled local authority spending to be strictly controlled from the center.

A slight easing of controls on social spending occurred in the last year of the Labor government, permitting an increase of 0.9 percent overall (after discounting the switch from child tax allowances to child benefit, which is offset by an effective rise in taxation on families). Planned spending for the next year (1979-1980) allowed for virtually no growth, and included the centrally provided social security and health services in the cutbacks. Though the Tories' June 1979 budget made further immediate cuts in the housing and education program, raised prescription charges, and cut back on employment services for that year, the basic plans for this next squeeze on social spending had already been decided on by the outgoing Labor government. The outcome was zero change in real terms in social spending for 1978-1979 to 1979-1980.

Then in November 1979 and March 1980, the Thatcher government took two more swings of the axe against the welfare state and planned for a real fall in spending in 1980-1981. Much has recently been made of its failure to control public spending over this past year, but if any sector was out of control, it certainly was not social spending, which barely changed for the second year running. True it was planned to fall by around 1 percent, but the main increases reflected the catastrophic climb in unemployment during the year: Spending on all employment services (e.g., redundancy payments, special employment measures, Youth Opportunities Programmes) shot up by 41 percent, and the cost of all unemployment benefits by 31 percent. The overshoot in these two programs was counterbalanced by serious cutbacks in the local authority services — education, personal social services, and especially housing — which all fell roughly according to plan. Nevertheless, total public expenditure rose by 2.4 percent due to some belated U-turns in industrial support, rising interest charges, and planned expansion in the military and law and order services.

So, in November 1980 and March 1981 the government entered round three. As far as the social services are concerned there will be no change in total expenditure from 1980-1981 to 1981-1982. Once more, a large increase in social security (unemployment costs up by a further 29 percent), plus some growth in total NHS and employment

TABLE 3.1 Social Expenditure in the UK (percentage changes at constant 1980 prices)

Financial years:	1975/1976-1977/1978 (2 years)	1977/1978-1978/1979	1978/1979-1979/1980	1979/1980-1980/1981 (estimated)	1980/1981-1981/1982 (planned)	1982/1983-1983/1984 (planned – 2 yrs.)
Social security	8.6	9.1(3.2)[a]	2.5(-1.5)[a]	3.5	7.0	1.1
N.H.S.	2.1	2.7	-0.7	1.6	2.6	2.8
Personal social services	0.5	3.9	4.3	-6.2	-0.5	4.5
Education	-4.3	1.6	0.5	-3.5	-4.0	-4.0
Housing	-15.7	-6.6	3.8	-13.2	-23.8	-28.8
Employment	10.7	0.7	0.5	40.8	2.7	-28.2
TOTAL SOCIAL SERVICES	0.1	3.6	1.6	0.2	0.0	-3.1
TOTAL SOCIAL SERVICES (adjusted)[a,b]	(-0.4)	(0.9)	(0.0)			
TOTAL PUBLIC EXPENDITURE[c]	-7.4	5.1(3.8)	1.0(0.0)	2.4	0.3	-3.8

SOURCE: Public Expenditure White Paper, Cmnd 8175, March 1981.

[a]Adjusted for switch from family allowances and child tax allowances to child benefit in these years.
[b]Adjusted to include phasing out of food subsidies between 1975/1976 and 1978/1979.
[c]"Planning total" plus net debt interest.

services, will be financed by swinging cuts in housing and education. The planned cutback for the following two years of 3 percent appears to be based on still too optimistic forecasts of unemployment.

In conclusion, we have witnessed an overall stagnation of expenditure on the welfare state over the past five years in spite of the unplanned-for costs of unemployment. This stagnation followed a rise of 28 percent in real resources for the welfare state in the previous four years, 1971-1972 through 1975-1976. There can be no question that a decisive turnaround took place in government policy toward social expenditure in the mid-1970s.

The conservatives' intention to reduce progressively resources for the social services marks a new approach. Previous expenditure plans by the Labor government anticipated a further growth after the "temporary" cuts which were necessitated by the economic crisis. Nevertheless, it is important to stress that this quantitative attack on the welfare state was initiated by the Wilson/Callaghan administrations in 1975 — the Conservative policy is simply more (or rather less) of the same. In turning to the qualitative shifts in policy we see more clearly the distinctive ideological reversal championed by the Thatcher administration.

THE IDEOLOGICAL ATTACK

The Tory government has turned around public policy. Its changes include the following. Sitting council house tenants now have the right to buy their house at very large discounts, council rents are going up sharply (which will further "encourage" the sale of council houses), new housing programs are pared to the bone, and new controls are to be introduced over councils' direct works departments. In education the trend to comprehensives is being halted, overseas students fees are being raised to their full economic cost, plans to introduce student loans have been mooted, and the new assisted places scheme will provide more public money for private schools. And in social security, the 1980 budget marked the most significant attack on social rights since the war: Benefits to strikers' families have been cut by £12 a week, "short-term" benefits for the sick, unemployed, and disabled were cut by 5 percent in real terms in 1980-1981, earnings-related short-term benefits will be abolished from January 1982 (but not earnings-related contributions), and responsibility for payment of basic sick pay during the first eight weeks of sickness will devolve on employers from 1983.

Drawing together the threads of these policies, we see a major attempt to reprivatize parts of the welfare state (housing, education), to shift charges onto consumers (health, housing), and in social security explicit decisions to weaken the organized working class, to widen the gap between "deserving" and "undeserving" claimants, and to encourage work incentives (at the same time that some employment schemes are eliminated and unemployment will escalate). This is not to mention the encouragement of police and law and order services, or the overt attack on progressive education. It is difficult to avoid the conclusion that the welfare state is attacked by the new conservatism at least as much for ideological as for economic reasons. Some Tory policies (the assisted places scheme, possibly council houses sales) will *raise* net costs to the exchequer.

RESTRUCTURING THE WELFARE STATE UNDER THATCHER

In *The Political Economy of the Welfare State* (1979: 141) I argued:

> [There are] four ways amongst others in which the state, acting in the long-term interests of capital, may seek to restructure the welfare state at a time of economic crisis like the present: by adapting policies to secure more efficient reproduction of the labor force, by shifting emphasis to the social control of destabilising groups in society, by raising productivity within the social services and possibly by reprivatising parts of the welfare state.

But I also argued that this process of restructuring must be situated within an overall strategy for counteracting the crisis, and that there were, in Britain today, two alternative capitalist strategies available: the radical right strategy and the "corporatist" strategy. The Labor administrations of 1974-1979 practiced a degenerated form of corporatism increasingly watered down with various monetarist elements. In the 1979 election a clear choice was expressed (at least in the southern half of England) for an undiluted form of the right-wing strategy. But what are the crucial elements of the "new conservatism," and how have they affected the restructuring of the welfare state?

In answering this question I am drawing on some previous articles in *Marxism Today*. Peter Leonard (1979) has explored different ways in which this restructuring takes place under social democracy and under the "radical right." Together with earlier articles by Stuart Hall (1979) and Andrew Gamble (1979), he emphasizes the specific role of political and ideological elements in this particular response to the contemporary crisis of British capitalism. I wish to consider the links between these and the Tories' economic strategy analyzed by Michael Bleaney (1979).

According to Hall and Gamble, Thatcherism is a political formation that combines the principles of the "social market economy"
with a new "authoritarian populism." The social market economy
represents a return to some of the precepts of nineteenth-century
liberalism: a limited role for government, an emphasis on the responsibilities of the individual, and so forth. In Britain and in the Anglo-
Saxon world generally this has taken the form of a resurgence of
monetarism as advocated by Milton Friedman.

Ideologically, Thatcherism marks a shift from the consensus politics which have characterized postwar British governments, both
Labor and Tory. It is (1) anti-trade union, attacking the organized
Labor movement on the economic, legal, and social fronts; and (2)
anti-poor, introducing soon after its election victory a budget that
reduced income tax rates notably for top income groups and raised
VAT to 15 percent. Within low-income groups, it (3) consistently
discriminates against the "undeserving" (i.e., working-age) poor, by
comparison with the elderly for example; it is (4) anti-deviant, em-
phazing the themes of law and order, e.g., extending punitive treatment for juvenile offenders; and (5) covertly racist, for example in the
new Nationality Act. Finally, Thatcherism is (6) individualist, inter-
preting social problems in terms of individual and family pathology, as
for example Sir Keith Joseph's emphasis on "problem families" and
the "cycle of deprivation": and (7) anti-state, constantly stressing the
dependence of the "wealth consumers" in the state on the "wealth
producers" in the private sector, and rejecting social cost-benefit
accounting.

The term "authoritarian populism" encapsulates many of these
distinct ideological elements. *Populism* here represents an appeal to
national interests which supposedly transcend sectional interests.
The prime target is of course the trade union movement, but it
encompasses all other interest groups representing the under-
privileged, such as racial organizations and many quasi-governmental
bodies (Quangos) such as the Supplementary Benefits Commission,
now abolished. All are seen to interfere with the operation of market
forces which supposedly work in a neutral way to augment national
welfare. It is an *authoritarian* populism, first because in many fields
the postwar shift to more liberal, humane policies is rejected. And
second, because paradoxically the new right requires a *stronger* state
to represent national interests and override sectional interests.
Crudely speaking then Thatcherism = monetarism + authoritarian
populism, though the economic and ideological components clearly
complement each other in many respects.

THE CENTRALITY OF THE WELFARE STATE

What is striking for our purposes is the position of the welfare state at the heart of these two strands. The welfare state is the central target for the radical right on both counts. First, because it allegedly generates even higher tax levels, budget deficits, disincentives to work and save, and a bloated class of unproductive workers. Second, because it encourages "soft" attitudes toward crime, immigrants, the idle, the feckless, strikers, the sexually aberrant, and so forth. Economic prescriptions and populist incantations are harnessed together, and their prime target is the expanded sphere of state responsibility, state regulation, and state-provided benefits which constitute the modern welfare state. As Hall (1979) stresses, this new-right ideology did not appear out of thin air; it needed to be constructed, though it utilized existing elements. And, as Gamble (1979) shows, it had to be welded into a party political program which could be electorally successful.

The process of restructuring the welfare state can now be situated within the political formation of the radical right. First, the *quantitative* role of cuts follows from the precepts of monetarism: strict control of the money supply, a substantial reduction in the level of government expenditure and taxation, and a shift toward indirect taxation. A reduction in the public sector borrowing requirement is a key object of policy because of its impact on the money supply (or on interest rates if government securities are to be sold to the nonbanking sector). Given the commitment to lower tax levels in order to encourage incentives to work and invest, this must involve even faster cuts in public spending. Given the commitment to higher defense spending, this must involve still greater cuts in social and economic expenditures. The ensuing policies weaken the power of organized labor via higher rates of unemployment (and the definition of higher levels of wage claims as "excessive"). The goal is to use market forces (together with new legal restrictions) to reduce real wages and augment profits. A cut in the "social wage," for example, reducing housing subsidies or personal social services, augments this pressure to reduce labor's share in the national income. It thus provides an indirect route to encourage profitability and reinvestment in British industry, even if a sound monetarist government like the present one disclaims any responsibility for the national rate of economic growth.

Second, the *qualitative* shifts in social policy are designed to reassert individualism, self-reliance, and family responsibility, and to reverse the collective social provision of the postwar era. Present attempts to impose a national curriculum and "raise standards" in public education provide a striking example of the social program of the new conservatism. In many ways, though not all, these qualitative shifts complement the absolute cuts in expenditure: Cutting social benefits to working-age adults saves money and panders to the anti-scrounger mentality of the new populism. Together these two sets of forces have generated the most sustained attack on the welfare state since World War II. The restructuring of the welfare state has begun in earnest.

WELFARE UNDER CAPITALISM

To understand why the welfare state is under attack today, we must first understand why it developed so markedly this century and in particular since World War II. This section summarizes the analysis developed at greater length in my book (1979). The welfare state in modern capitalist countries comprises public cash benefits (such as pensions), public benefits in kind (such as education), public regulation of the activities of individuals and corporations (such as consumer or labor legislation), and the taxation system. This complex of measures serves two major goals: the reproduction of labor-power, and the maintenance and control of the nonworking population. Both involve quantitative and qualitative aspects; the state ensures directly or indirectly a minimum level of consumption for different groups in the population, and at the same time, it modifies the pattern of socialization behavior, and abilities within the population. Since it is a capitalist welfare state, it imposes sanctions and controls at the same time that it provides benefits (take council housing, for example, or education, or supplementary benefit).

These goals can be grouped together and constitute *the state organization of social reproduction*. As Engels pointed out, reproduction is just as essential an activity in all societies as production. The process of social reproduction refers to the processes of biological reproduction, of economic consumption, and of socialization: in short, to the way individuals as social beings are "produced" rather than goods and services in the process of production. Of course the family has played, and continues to play, a crucial function here, but when we refer to the state organization of social reproduction we are referring to the way the welfare state has modified and partially supplemented the family in the twentieth century. From supplementary benefit to child care officers the welfare state today is intimately involved in this process. But our definition also reminds us how contentious this process is: After all, the radical right is questioning precisely the respective roles of state, family, and market in the sphere of social reproduction.

Reproduction is only one of the activities of the modern capitalist state. Modifying James O'Connor's (1973) analysis, we can identify three broad functions that it performs:

(1) accumulation
(2) reproduction
(3) legitimation/repression

The first refers to all the means by which the state tries to maintain favorable conditions for the accumulation of capital. The third refers to the means by which it attempts to maintain social order and social control while at the same time trying to preserve social harmony and avoid harmful conflict (harmful, that is, to the state and private capital). Social policies clearly have implications for these other goals of the modern state: For example, some policies (like the redundancy payments scheme or parts of higher education) are designed to achieve economic benefits for the private sector, while others (such as some aspects of education) perform a social control job, and many help to legitimize the system and reduce dissension in society. So the welfare state is involved in all three areas of activity, though I would argue its prime concern is with maintaining and adapting social reproduction.

THE WELFARE STATE: THE OUTCOME OF CONFLICT

This scheme does no more than provide a framework for understanding particular social policies such as those of the Thatcher government. In fact it scarcely does this, for there is a danger in the above account of *explaining* the modern welfare state in terms of the *functions* that it performs in capitalist society — the danger, in short, of a crudely functionalist interpretation. But class conflict, in particular, pressure from the organized working class, has played a major role in the origins and spread of welfare services. Indeed, for many the British welfare state is the child of the labor movement and the postwar Labor government. How can this view be reconciled with the role that, we have argued, it plays in securing capital accumulation, social reproduction, and political legitimation?

The brief answer is that the welfare state is the vector of two sets of political forces: "pressure from below" and "reform from above." The first refers to the myriad ways in which class movements together with social and community movements demand social reforms to protect or extend their interests. This may result from pressure group politics within the state at one extreme to direct action and street conflict at the other. "Reform from above" refers to the various ways in which the state seeks to implement social reform which will serve the longer-term economic, social, and political interests of capital, or particular sections of capital. The state does not automatically perform this job: It requires at the least the executive and administrative wherewithall and a form of political mobilization. I believe these are more readily available the more centralized is the apparatus of the state. The stronger is the role of the executive and senior civil service vis-á-vis parliament, for example, the more can they override short-term pressures from representatives of particular capitalist interests and impose a longer-term, more class-oriented strategy.

I would argue that these two sets of forces have both strengthened, and partially reinforced one another, over the last forty years, particularly in Britain. Their interaction led to two especially notable periods of social reform in the 1940s and the 1960s. During World War II, the foundations of the Keynesian-welfare state were established. Though many of the reforms were enacted by the postwar Attlee government, it is notable that most were the product of the wartime coalition government (the Beveridge report, the Butler Education Act). They represented the outcome of pressures from below—wartime radicalization and a spirit of "no going back to the Thirties" — plus reforming drives from above — a recognition of the political and economic necessity of greater state responsibility for economic performance, notably through Keynesian demand management techniques. Thus, in Britain Keynes and Beveridge represented a linked response to the prewar crisis focusing respectively on demand management (part, but only part, of the economic sphere) and the sphere of social reproduction. Together they formed the core of the "post-war settlement" between capital and labor which was to prove so successful a basis for postwar prosperity.

THE SIXTIES AND SEVENTIES

The 1960s and early 1970s saw a second wave of reforms as social and economic policy was slowly restructured in the face of a faltering rate of accumulation in Britain. The development of incomes policies, industrial strategies, the modification of Beveridge's social security principles, the expansion of higher education — these all represented a further extension of state intervention and a closer gearing of economic and social policy. Again there were pressures from below, from a labor movement of greater defensive economic strength, and from newer social and community movements; and pressures from above as reformers and spokesmen for capital understood the limited role of Keynesian policies and advocated more systematic economic and social intervention in order to restructure British capital in the face of overseas competition, domestic class pressures, and a falling rate of profit. Again, though the Labor election victory in 1964 signalled the shift in direction, many of the individual reforms had been initiated in the last years of the previous Conservative administration.

So the British welfare state represents the outcome of growing working-class pressure for economic and social reforms, modified by the desire of a more centralized state apparatus to restructure economic and social policies for its own reasons. In part the postwar Keynesian welfare state generated its own momentum for further state intervention, to secure economic growth and capital accumulation within a new balance of class forces that it had itself helped to shape. It follows from this that the development of Keynesian economic policy and modern social policy were interlinked and formed the two central planks of the postwar political consensus between the parties. It therefore comes as no surprise to find that both planks are simultaneously under attack from the new radical right.

THE WELFARE STATE AND THE ECONOMIC CRISIS

What is the link, if there is one, between an expanded welfare state and a declining economy? Is the British economic crisis the result of an overgrown public sector, as the present government would have it, or are the two unconnected? My own view is that there is a link, but that it is not so straightforward or unambiguous as the new conservatism suggests. After all, a recent European Economic Community (EEC) report showed that government spending as a share of GNP is lower in the United Kingdom than any other country in the EEC. On the other hand, socialist reformers and others who deny any link and who reiterate Keynesian nostrums about the need for more public spending in order to pull us out of recession do a disservice to the socialist movement. The Keynesian welfare state *has* generated new contradictions, working as it does within the framework of a private capitalist economy. It is not possible for state spending to rise inexorably as a share of GNP without adverse consequences for its domestic capital. What then are these limits?

I believe there are two main limits. First, a growing level of state expenditure exacerbates the postwar conflict between capital and labor over the *distribution* of national output. Given the centralization of capital within an expanding international economy, and given a stronger, more organized labor movement, inflation becomes built into advanced capitalist countries, as capital and labor can in turn offset higher wage costs or higher prices. When the state then lays claim to a higher share of resources, this two-way conflict becomes a three-way conflict adding to the inflationary pressure. For however the state seeks to finance this expenditure — via higher taxes on the working class, or on corporations (very unusual), or via higher indirect taxes, or via higher state borrowing — the result is to exacerbate the spiral of wage costs and prices. At a time of economic slowdown and increasing international competition, this has also contributed to the squeeze on profit rates.

Second, the growing level of state expenditure and intervention interferes with the *production* of surplus value and profit.[4] The growth of the "social wage" and "collective consumption" means that the operation of the labor market and the reserve army of labor is impaired, and the bargaining strength of the working class increased. Unemployment benefits, family benefits, public assistance, state health and social services, housing subsidies, and so on all remove part of the real living standards of the working class from the wage system, and allocate this part according to some criteria of social need and citizenship. Citizenship rights are counterposed to property rights, and the ability of capital to transform labor power into labor performed is impaired. Now the precise impact of welfare policies on capital accumulation will depend in practice on the criteria according to which social benefits are distributed, and will vary between countries. Are they predominantly distributed on some criterion of need, or do they take into account the labor market position of men and women, the impact on work incentives, and so forth? In other words, the extent to which particular social policies *subvert* or *complement* the market mechanism can only be answered after detailed research of the way in which they operate. Generally speaking, one may assume that the impact of welfare policies on capital accumulation will be more favorable the more they follow market criteria in distributing and awarding benefits, and the more closely social and economic policies are integrated. But of course, the further social policies are shifted in this direction, the more they may interfere with social reproduction and political legitimation.

In the light of this, what has been the impact of the welfare state on British capitalism? Two peculiar facts about Britain must be borne in mind in answering this question. First, the position of Britain within the world economy is declining, and the deep-seated weaknesses of our economic structure are now superimposed on a worldwide recession which has marked the end of the postwar boom. Second, the defensive economic strength of the British trade union movement has prevented the strategy of industrial restructuring attempted by Labor and Conservative administrations since the early 1960s from being successfully implemented (Purdy, 1976; Jacques, 1979). This defensively strong, decentralized labor movement with extensive shop-floor organization has also hindered the restructuring of the welfare state. Unlike the United States, Britain has a developed set of social services available in the main for the whole population rather than certain privileged strata within it. But unlike Sweden and West Germany, for example, these are not closely integrated with economic policy to achieve greater mobility of labor or to encourage labor force participation. Council housing policy, for example, may well restrict the mobility of labor demanded by capital and interfere with the operation of labor markets. Furthermore, British unions managed to maintain their members' posttax incomes in the face of slow growth, rising tax levels, and periodic incomes policies at least until the mid-1970s. But again, unlike Sweden this was not achieved by means of a corporatist-style social contract which would yield some tangible benefits to capital. In an environment of relative economic decline, it is likely that the British welfare system has contributed to the British economic crisis by exacerbating inflation and undermining market mechanisms. These very features stem from the particular ways social policy has developed in postwar Britain, outlined above.

It is perhaps not surprising, then, that monetarist and populist attacks on the welfare state have established themselves here. Given the failure of Keynesianism, and the progressive degeneration of the Labor governments' corporatist experiments after the 1974 social contract, a vacuum opened up which first Powell, then Thatcher, Joseph et al. were quick to exploit. The defeat of the left in the EEC referendum of 1975 under a Labor government helped prepare the ground for this move to the right. The indigenous populist ideology, the reluctance of British capital to opt for Continental-style interventionism, and the failure of the Labor leadership to develop an alternative strategy to replace the wilting nostrums of Keynesianism — all left the way open for a tax-welfare backlash culminating in the victory of the new conservatism in the last election. The left is also implicated in its failure to unify around some coherent transitional strategy as an alternative to both Labor reformism and the new conservatism. The alternative economic strategy adopted at the Labor party conference in 1973 represented an important stage in the process of rethinking. But as Hodgson (1979) has argued, it was conceived by the left of the Labor party, in particular the Tribune group, as a parliamentary policy without the need for mobilization at the local level. As a result, its socialist content was weakened to the point where it differed little from corporatist-style interventionism.

The outcome of these events is that Britain is experiencing the most far-reaching experiment in new right politics in the Western world. The situation is rooted in the contradictions engendered by the extension of citizenship rights and a developed welfare state within a declining capitalist economy. The two principal contradictions are the exacerbation of class conflict over the distribution of the national product, and the undermining of the production of the surplus product. But if the Keynesian welfare state helped sustain the economic and social relations of advanced capitalism, we would expect that its erosion would generate new contradictions. In fact this is exactly what we find.

THE CONTRADICTIONS OF THATCHERISM

Let us consider each of the three functions of the state mentioned earlier on.

Accumulation. All economic forecasters now agree that the prospects for the British economy at least until the end of 1983 are grim. Unemployment of about 2½ million (March 1981) must now rise to 3 million and beyond. In the single month of December 1980, the equivalent of Bradford's entire workforce was added to Britain's dole queues. Manufacturing output has fallen an unprecedented amount in this century — 15 percent since late 1979. The government expects that 1981, like 1980, will see a 2-2.5 percent fall in real gross domestic product. The Cambridge Economic Policy Group forecasts are consistently worse than these official forecasts. Thus, the deflationary monetary policies of the present government hardly provide the basis for a recovery in profitability and capital accumulation in the short-to-medium term: Instead, they are making matters much worse.

The cuts in social expenditure will have an impact not only on the recipients of the services, but on sections of private capital as well. So far it is capital spending that has borne the brunt of the cuts, so that government demand for the output of the construction industry, for example, has fallen dramatically. The new cuts announced recently will only exacerbate this problem. One half of National Health Service (NHS) expenditure, for example, consists of purchases from the private and nationalized sectors (drugs, supplies, oil, electricity), so that government cuts here directly harm private industry. Even if the government manages to impose its future cuts on transfer payments, public sector employment, and wage and salary levels, this will still have a multiplier effect on private sector output.

How can it happen that a policy intentionally designed to revitalize British capitalism should have this opposite effect? The answer may be found in a contradiction of the accumulation process long ago noted by Marx. For capital accumulation to proceed two conditions must be fulfilled: First, profits must be produced by successfully exploiting labor within the production process; second, these profits must then be realized by the exchange of commodities in the circulation process. These two moments of the process are now in contradiction. Keynesian policies overcame the interwar crisis of underconsumption by sustaining aggregate demand, but over the long term they have contributed to a falling rate of profit and undermined the production of surplus value. In reaction, the present government is attempting to alter the class balance of forces to raise the rate of exploitation; in so doing, however, it will worsen the domestic conditions for realizing surplus value in the short term. That is, a deflationary policy results in excess capacity and falling profit margins (unless exports can rise to make up the loss, a scenario that looks increasingly unlikely). Thus, higher government spending facilitates the realization of profits in the short term, but interferes with the production of profits in the longer term, whereas cuts in spending may aid the production of profits in the longer term, but at the short-term expense of realization. The conclusion is that an approach relying on market forces to expand accumulation in the longer term worsens its prospects in the short to medium term.

Reproduction. The new conservatism believes that the welfare state, the state organization of social reproduction, has proceeded too far, that individual and family responsibility have been undermined and need to be restored. But to what extent can this twentieth-century process be put into reverse? In my view, the problems generated by a substantial dismantling of welfare services would be great, a fact reflected in the hesitancy even of this government to implement spending cuts in, for example, the health services. First, it would throw a greater burden back on individual families and in particular, on the women within them. (Moroney [1976] documents the enormous burden borne by, for example, middle-aged women in caring for their elderly relatives.) But rising numbers of conventional nuclear

families are breaking up in divorce, and a growing proportion of them
are not being reconstituted. The number of children living with single
parents is increasing, adding to the demand for supplementary benefits and social services. Furthermore, the women's movement is
now a force capable of resisting some of the more overt attempts at a
"back to the family" approach, as its recent success in deflecting
attempts to retract the abortion laws testifies. Second, social needs
are not something objectively identifiable; they are interpreted and
new needs engendered in the process of class and social conflict.
Community-based movements such as Womens Aid or law centers
have helped recognize and define previously personal problems as
new social needs. So, too, have state social services, as when the
introduction of the National Health Service in 1948 revealed a large
unmet need for medical services. These discoveries are not easily
unlearned and the gains in social provision not easily reversed.

Thatcherism, in attacking the Keynesian-welfare state couplet,
thus risks regenerating many of the problems these policies were
initially designed to overcome: mass unemployment, renewed poverty (particularly among children), uneven regional development, and
urban decay among others. But it will be attempting to reverse a
process that has generated an entirely different environment to that of
the thirties. The changes are critical: Expectations for social provision now exist among a majority of the population; and new movements exist, premised on the welfare state, to extend and defend
existing social and community provision.

Legitimation/repression: a hazardous operation. The present
government seeks legitimacy for its policies in the ideologies of
economic realism and authoritarian populism, but these may prove to
be a fragile basis, for several reasons. First, the purely electoral
consequences are hazardous because, as we noted above, the immediate impact of the policies is to worsen recession and unemployment and to lower real incomes. Some monetarists appreciate that the
long timespan of their policies conflicts with the election cycle of
liberal democracies, and either bemoan the constraints this imposes
on "sound" government policies (something missing after all in the
Chilean monetarist experiment) or conduct a vigorous ideological
offensive to convince the electorate that theirs is the only sure way
forward. But pressure is building up on the Thatcher government
from within the Conservative party to moderate its monetarist zeal, if
not yet to undertake a U-turn.

Second, one fifth of the labor force is employed by central and
local government and has a direct interest in maintaining the level of

public expenditure and employment. Furthermore, a majority of the population are consumers in one way or another of social services, and many will resent falling standards in their particular sector, even if supporting the government's broad objectives. Third, fears are growing (witness the postmortems held on the Bristol riot) about the threats to law and order and the growth of widening social divisions and conflict which present social policies will exacerbate. Growing numbers of unemployed school leavers, or the ghettoization of the council house estates remaining after the sale of the better local authority housing, are two examples of the political dangers in dismantling socially provided services too far. It is true that the government is developing a more repressive strategy in some areas of the welfare services and in its law and order policies generally, but the harm to its legitimacy should not be underestimated, particularly if major opposition to its policies develops among those most affected by them. Lastly, the attempt by the Thatcher government deliberately to depoliticize areas of social life by disclaiming government responsibility and returning them to the anonymity of the market may itself be politically unacceptable. After several decades in which the state has accepted responsibility for the rate of unemployment, it is hazardous for it now to claim that it has been powerless all along.

POSTSCRIPT: TOWARD A LEFT ALTERNATIVE

Since early 1980, events have lent support to this analysis. But a coherent opposition to the politics of the Thatcher administration has yet to arise. A tentative balance sheet of the opposition by specific movements would look something like this.[5]

(1) The public sector unions until 1981 maintained and even increased the real wages of their members, but have had little success in preventing cuts in services and employment (with significant exceptions, such as the miners).

(2) Social movements of clients, women, and other community groups have organized opposition to government policies at the local level, but aside from some specific successes (e.g., preventing ward closures in some hospitals, preventing the watering down of abortion services) the overall record is not good.

(3) Some Labor-controlled local councils, e.g., in parts of London and Yorkshire, have defied government cuts and raised their rate income to maintain services, but this loophole is now to be blocked by central government in a move which will weaken

still more the autonomy of British local authorities and their
powers of resistance.

In the absence of a unified political resistance, what has occurred
are spontaneous riots by black and white youth in some cities. In most
cases this is directed against police repression, "saturation policing,"
and so on, but many participants are quite explicit about the lack of
amenities and opportunities open to the poor and oppressed groups in
Thatcherite Britain. To take one example, one in six are underemployed in Liverpool today, in the Toxteth district it is over two in
five, and among young people in Toxteth it is nearer three in five.

Ultimately, however, the future of both Thatcherism and the welfare state will depend on the *political* alternatives available. At the
time of writing, the political viability of a new center coalition of the
Liberals and Social Democrats is unclear. Just as unclear is the
strategy they would offer in the field of economic and social policy to
tackle Britain's deep-seated crisis. Attempts to decry the usefulness
of any strategic approach at a time of great uncertainty, and to uphold
the virtues of pragmatism, will soon run foul of the contradictions
noted above, though North Sea oil may provide a temporary buffer.
Given the bankruptcy of Keynesian/welfare state reformism, the
alternative to Thatcherism offered by a new center coalition would
most likely be a further step forward in state centralization and state
intervention, obscured by a rhetoric of decentralization.

It is important to recognize that there is nothing inherently
socialist in further state intervention, and that either option will
threaten some of the political and social rights established since the
war. It is also important in opposing both strategies to recognize the
contradictory nature of the contemporary welfare state: It signals a
collective responsibility for meeting an array of social problems and
social needs certainly, but it achieves this through a process of centralization in which social policies are deformed and adapted to suit
the requirements of capital and to minimize democratic control.

The need is to move beyond the traditional politics of the Labor
party by harnessing the labor movement to the social, community,
women's, and other democratic movements that have partly developed in and against the welfare state. Second, it is necessary to
combine this with an alternative economic strategy to be implemented by a future left government at the national level. An urgent
task in both respects is to develop a parallel alternative *social strategy*
that will propose new priorities for social policy together with new
forms of implementing and controlling it. Insofar as this is not

achieved, the Thatcherist strategy, despite its problems outlined above, could win by default.

In all countries presently afflicted with the politics of the new right, it is vital to avoid the temptation to seek refuge in the politics of the middle way, of liberal Keynesianism and social democratic reformism. As I have argued, neoconservatism is in some respects a genuine response to the contradictions of this phase of capitalism in the advanced western countries. The need to construct a "third alternative" is now urgent. It is a time of great danger but also of great potential, since in the face of the bankruptcy of the ideas and values of the two existing political currents, socialist ideas and values may secure a purchase in the heartlands of capitalism.

NOTES

1. This is a modified and updated version of that paper.

2. It remains to be seen to what extent the Mitterand regime in France pursues a centrist policy or an "alternative left" policy, or how the two elements are mixed. It does, however, inherit an extremely strong, if inegalitarian, capitalist economy from its right-wing predecessors (The Economist, October 11, 1980).

3. For a recent comparison of the redistributive effects of the welfare systems of Britain, the United States, Sweden and France, see Stephens (1979: ch. 5).

4. This second limit imposed by the welfare state on capital accumulation is overlooked in my book (1979). Its importance is well argued by Harrison (1980) in a review. See also Bowles and Gintis (n.d.).

5. This clearly needs a lot more research. For an account of the opposition to Conservative cuts in the London borough of Wandsworth, see the first issue of the new British journal, Critical Social Policy, May 1981.

REFERENCES

BLEANEY, M. (1979) "The Tories' economic strategy." Marxism Today (October).
BOWLES, S. and H. GINTIS (n.d.) "The crisis of liberal-democratic capitalism." (unpublished).
GAMBLE, A. (1979) "The decline of the Conservative Party." Marxism Today (November).
GOUGH, I. (1980) "Thatcherism and the welfare state." Marxism Today (July).
——— (1979) The Political Economy of the Welfare State. London: Macmillan.
HALL, S. (1979) "The great moving right show." Marxism Today (January).
HARRISON, J. (1980) "State expenditure and capital." Cambridge Journal of Economics 4, 4.

HEAP. S. H. (1980-1981) "World profitability crisis in the 1970s: some empirical evidence." Capital and Class, No. 11.

HODGSON, G. (1979) Socialist Economic Strategy. London: ILP Square One Publications.

JACQUES, M. (1979) "Thatcherism: the impasse broken?" Marxism Today (October).

LEONARD, P. (1979) "Restructuring the welfare state: from social democracy to the radical right." Marxism Today (December).

MORONEY, R. (1976) The Family and the State. London: Longman.

O'CONNOR, J. (1973) The Fiscal Crisis of the State. New York: St. Martin's.

PURDY, D. (1976) "British capitalism since the war." Marxism Today (September).

STEPHENS, J. D. (1979) The Transition from Capitalism to Socialism. London: Macmillan.

4

Social Services Under Reagan and Thatcher

PAUL ADAMS

GARY FREEMAN

☐ NUMEROUS COMPARISONS BETWEEN the fortunes of Margaret Thatcher's Tory government in Britain and Ronald Reagan's Republican regime in the United States have appeared recently in the popular press. Many observers have noted the striking similarities between the two administrations and have sought, especially on this side of the Atlantic, to develop cautionary lessons for the Republicans on the basis of the tribulations of Thatcher. This sort of comparative analysis is not limited to outsiders, either. Shortly before the newly elected president assumed office, two of his key advisors circulated a memorandum, based in large part on an exploration of Thatcher's experience, which urged Reagan to avoid "an economic Dunkirk" by moving swiftly to reduce federal spending, cut taxes, and control the money supply (Wall Street Journal, December 18, 1980).

Reagan's program is indeed similar to Thatcher's, and the problems she has faced are likely to trouble him too. Economic conditions in the two countries, and the political context within which the current experiments are being carried out, are, however, sufficiently disparate to suggest the need for caution in drawing inferences from the record of one for the behavior of the other. Furthermore, the actual policies being enacted in some cases bear only a tenuous relationship to the more general rhetoric.

The ultimate success of Thatcher's and Reagan's global policies depends on the achievement of two separable but interdependent outcomes. The first is that reforms derived from the logic of the overall economic and social strategy must be implemented in specific program areas. Implementation is the precondition for success, and in each instance the particular political coalitions and strategic opportunities associated with individual programs will determine the extent to which it is accomplished. It is only when the reforms dictated by the strategy are in place that the second and decisive stage of the process is reached, the period in which the legitimacy and utility of the plan is validated by bringing about the predicted economic revitalization (measured, to be sure, by constantly shifting criteria) or is discredited for not doing so. The programs can fail, then, at either stage, but the meaning of failure will be very different in each case.

This chapter attempts to situate the Thatcher and Reagan policies toward one set of programs — the personal social services — within their more general political and economic strategies. The social services with which we are primarily concerned are those that involve subcentral governments in a fiscal relationship with central government and that are delivered by social service workers. We assess the relative success of the two governments in implementing social service reductions and reorganizations and discuss some of the likely consequences of the reforms once they are set in motion. Our analysis is constrained by the comparative youth of the Reagan administration, which requires us to limit our comments primarily to his proposals, while in the case of Thatcher we can discuss both the problems she has encountered implementing her plan and some early indications of its consequences. We do not claim that the social services are representative of other categories of state activity. On the contrary, the evidence suggests that in both countries they are among the weakest and most vulnerable targets of the budget cutters and among the most amenable to retrenchment. We endeavor to show, nevertheless, that the prospects that deep cuts in social services will contribute significantly to a reduction in state spending as a whole or to economic recovery are more apparent than real.

THE CUTS

In Britain, as in the United States, the present effort to reduce state spending on the social services represents in part simply an escalation of the policies of the preceding administration. The potential demand for social services has been growing rapidly, as the numbers of single-parent families (which doubled in the 1970s), children in care of local authorities, and elderly, especially frail elderly, increases far more rapidly than the general population (Townsend, 1979: 902-912). The existing broad responsibility of local governments to provide social services was extended in the early and mid-1970s when certain programs for the social care of the chronically sick, disabled, and elderly were transferred from the National Health Service to the local authorities. The combination of growing demand and increasing programmatic jurisdiction contributed to a rapid expansion in spending for the social services by local governments and to a subsequent increase in central government grants for the purpose. Ken Judge has estimated that total spending on the personal social services grew by 219 percent between 1955 and 1975 (in constant prices) and that the average annual real increase between 1969-1975 was 12.7 percent (1978: 1-4).

The response of the 1974-1979 Labor government to these pressures was to make drastic cuts in capital expenditures and then to try to hold down real growth in social services by cutting central government subsidies to local authorities and by imposing cash limits on their expenditure growth. The cash limits system was an especially draconian means to prevent price rises and unanticipated pay increases from pushing government expenditure beyond budget estimates. First employed with respect to local government building programs in 1974-1975 and 1975-1976, they were introduced in a systematic fashion in 1976-1977 (Cmnd. 6440, 1976). Despite these efforts, during the period from fiscal year 1973-1974 to 1978-1979, every category of social welfare expenditure grew except education, and spending for the personal social services went up by 15.9 percent (Judge, 1980: Table 4).

Where Labor failed, the Tory government of Margaret Thatcher intends to succeed. In its 1980 White Paper on public expenditure, the government firmly stated its intention "to reduce public expenditure progressively in volume terms over the next four years . . . . [and] to give priority to defence, and law and order" (Cmnd. 7841, 1980: 3, 6). The principal programs targeted for cuts are industry, energy, trade, employment, housing, education, and net borrowing by nationalized industries. Spending on housing, for example, is slated to suffer a fall of 47 percent between 1980-1981 and 1983-1984. Expenditure on education, science, the arts, and libraries will be reduced from £9.2 billion in 1980-1981 to £8.7 billion in 1983-1984 (in constant prices). On the other hand, outlays for defense and law and order will rise by £743 million and £170 million, respectively, during the same period. In spite of its determination to bring about a shift in national spending priorities, the government will spare some social programs absolute cuts in expenditure, though not in the services they will be able to provide. Social security, for instance, is projected to grow by about 7.3 percent. It is much the same with the personal social services. The White Paper projects total central and local spending for 1980-1981 at a figure 4 percent below the level for 1978-1979. After that, however, current spending is expected to rise at an annual rate of 2 percent while capital spending is held constant. This will mean, nonetheless, that there will have been, in a period of rapidly rising demand and costs, a drop of 14 percentage points in the rate of growth of total spending on the social services between 1974 and 1984 (Judge, 1980: Table 4).

The squeeze comes from the national government, but it is at the level of the "local state" that decisions are made about what services will be eliminated or reduced in order to stay within centrally mandated limits. The government's spending plans assume a modest increase in demand for services but deliberately ignore the problem of rising costs. The purpose of the cash limit system is to "require spending Departments and authorities, in managing their expenditure, to provide within these limits for any increases in costs due to pay or price increases" (Cmnd. 6440, 1976: 5). There is considerable variation among local authorities in this respect. Many have increased local taxes to offset part of the loss of national revenues, but none has escaped the necessity of reducing the quantity or quality of services for the elderly, children, the disabled, the mentally ill, and retarded. They have imposed or increased charges for many services, such as meals-on-wheels and home helps, and have lowered levels of staffing, both directly and by keeping positions unfilled. They have closed residential homes ahead of schedule while postponing the provision of day-care centers and other community services (New Society, July 10, 1980).

"Vulnerable" clients may suffer not only the loss or increased cost of several of these services, but also the impact of cutbacks in other areas of the welfare state, such as education and housing. They are also among the hardest hit by the industrial stagnation, the highest levels of unemployment in half a century, and the double-digit inflation which have been the early fruits of Thatcher's economic policies. At the same time, the capacity of private, informal, and voluntary efforts to substitute for public provision, as conservatives would like, is much in doubt. As women become an increasingly important and permanent part of the workforce in expanding sectors of the economy, the unpaid female labor on which traditional, informal, and voluntary efforts depend becomes less available, whether in the form of family foster care or the home nursing of frail elderly parents. Organized voluntary efforts are highly dependent upon public funds and show little sign of substituting for them. They will be hurt not only by loss of government grants and purchase-of-service contracts, but also by the discouraging effects of marginal tax rate reductions on private charitable giving. Where private services *are* growing as an alternative to declining public provision, as in the health sector, the poor and vulnerable clients of the social services are the least likely to benefit.

In the United States, too, after a period of rapid expansion, there was a concerted effort to trim social service funding for several years prior to the advent of the Reagan administration. The federal government significantly expanded its support to social services in 1962. Amendments to the Social Security Act in that and subsequent years stimulated, partly by design and partly by accident, an explosion of spending for social services by the states and a concomitant increase in federal matching grants (Derthick, 1975; Richan, 1981: 151-179). From 1963 to 1966 federal social service expenditures rose an average of 22.7 percent each year. In the next four years they shot up at an annual average rate of 70.2 percent. By 1972 Congress had acted to put a $2.5 billion lid on funding for social services, and it reorganized the federal role in social service provision in 1974 with the enactment of Title XX of the Social Security Act (Richan, 1981: 176). Jimmy Carter's last budget was a treatise on the necessity of reducing social spending and the importance of attaining mastery over those items that had come to be labeled "uncontrollables," among which were social service matching grants, and that were estimated to constitute 75 percent of all federal outlays in 1980 (Wall Street Journal, January 16, 1981).

Ronald Reagan's proposals involved major reductions in nonmilitary public expenditures and a reorganization of the fiscal relationship between Washington and the state and local governments. This was to be accomplished through the creation of a number of block grants into which a large number of categorical programs would be collapsed. Overall, federal grants to states were to be cut by about 25 percent.

As the budget has emerged from Congress, it is exceedingly close to the President's figures, though the block grant scheme has run into some predictable problems. The budget proposes to slash some $35.2 billion from domestic programs. Among the hardest hit are elementary and secondary education (down about 20-25 percent), food stamps (down $1.66 billion out of $12.6 billion), child nutrition (down $1.5 billion), Medicaid (down $1 billion over the next four years), the Comprehensive Employment and Training Act (the public service component of CETA is eliminated, the rest is cut 20 percent), child abuse programs (down from $27.9 million to $19 million), and subsidized housing (down $12 billion) (New York Times, August 2, 1981).

Congress altered the block grant proposal in a number of ways: It created one block grant of $598 million for education programs, but most items affecting the needy retain their categorical status. In the health field, three block grants were developed with funding cut by around 25 percent. The social service block grant proposal died in conference committee. Reagan had asked that the Community Services Administration, foster care, adoption assistance, child abuse, programs for run-away youth, and child welfare be added to Title XX funds and be appropriated as a whole. Congress balked, however, but did reduce funding for Title XX from $3 billion in 1981 to $2.4 billion in 1982 (Congressional Quarterly Weekly Report, August 1981).

For a low-income family the combination of higher charges (e.g., for school meals and day care), lost eligibility (e.g., for CETA), and reduced or eliminated services (e.g., legal services and health clinics) will have a cumulative adverse effect on the quality of life. In the sphere of nutrition alone a family may suffer loss of the summer food program in its area, loss or reduction of food stamp benefits, and new or increased charges for meals at schools and day care centers. Substantial cuts in Medicaid, AFDC, and many other programs will severely limit the possibility of substituting other income to maintain nutritional levels.

THE STRATEGY

Cuts in social services are, of course, part of a larger economic and political strategy. At one level, the general economic plan is a technical exercise to reduce inflation and stimulate investment. Deflationary monetarist measures are applied (although the money supply, and inflation, at first increased dramatically under Thatcher) together with reflationary military spending increases and regressive "supply side" tax changes (cuts in the United States; in Britain, shifts from income to consumption taxes). Reagan's strategy differs from Thatcher's in its emphasis on supply-side stimuli and its lack of concern with "shaking out" or restructuring U.S. manufacturing industry to force out inefficient units. Profound cutbacks in nonmilitary government commitments are central to both strategies.

It is not our purpose to examine the economic basis of these measures, although their outcome (disinflation without deflation, or mass unemployment with high inflation and industrial stagnation) will be decisive for the legitimacy and survival of the administrations in question, together with their whole political programs. It is necessary however, to emphasize that these technical economic measures are not innocent in their social and political effects.

In the most general terms, both monetarist and supply-side elements of the strategy involve a redistribution from poor to rich, and from labor to capital. Monetary policy, it is true, may have devastating effects on business as well as labor — it has produced in Britain the highest failure rate of small businesses as well as the highest unemployment since the height of the Great Depression (Political Quarterly, 1981). But the supply-side tax changes deliberately widen the gap between rich and poor, aiming to shift resources from working-class social consumption to private capitalist investment.

Insofar as the net result is, as it has been to date in Britain, mass unemployment and weakening of the social protections of the welfare state, the strategy constitutes a sustained attack on the strength and bargaining power, as well as the living standards, of the working class. For Ralph Miliband, this "class war Conservatism" is the essence of Thatcher's strategy. "The enterprise consists," he argues, "in the radical reduction of the organized working class pressure to which governments have been subjected since 1945." It is a rejection of the traditional Tory policy of concessions to the working class and accommodation with its leaders, and a sharp break with the "social contract" approach to managing class conflict. It aims to weaken decisively the pressure from below exerted by the unions, their rank and file, the poor, and the lobbies and pressure groups that speak for them. In characterizing their task as being "to shift the balance as far as it will go in favour of managerial power," Miliband sees "militants, strikers, pickets, subversives" as the Thatcherite target: "For it is they who make greater productivity impossible, whose wage demands fuel inflation, who stand in the way of recovery, who ruin sound (i.e., toothless) trade unionism. This is the wisdom of outer suburbia; and it is in a commanding position at the centre of government" (Miliband, 1980: 278-279).

To this account Thatcher herself might reply, "Je ne suis pas thatcheriste," and it is true that Thatcher, like Reagan, rejects the earlier economic orthodoxy which blamed inflation on the cost-push of workers' wage demands. Both blame government spending instead, and oppose efforts to hold down wages directly through government controls, now the "liberal" alternative to their policies. But Miliband (1980: 279) both captures the spirit of class hatred which Thatcher draws on and represents, and points to the coherence of her *social* strategy, by comparison with which the Conservative economic strategy is a "rag-bag of hopes, hunches, prejudices and dogmas." Her social aim is to "produce a 'social climate' favourable to capitalist enterprise."

It is in the context of that goal that the cuts in social provisions (in particular, such measures as the reduction in benefits for families of strikers), as well as the generation of large-scale unemployment, are to be understood. Cuts in the social wage not only reduce public expenditure and with it, in the monetarist view, stagflation; they also make workers and their families more dependent on their direct, individual wages. Just as the Poor Law Reform of 1834 sought to head off the worker's retreat to the rural parish and indirectly reinforced managerial authority (Polanyi, 1944; Bendix, 1956), so Thatcher's social policy is directed at the strengthening of labor dicipline through the removal of social protection. The removal of such disincentives to work will, she hopes, cure the "why work syndrome," secure the primacy of the market as allocator of society's resources, and strengthen family responsibility and self-reliance.

Working-class consciousness, trade union strength, and the historic middle-class fear and resentment of them are less developed in the United States, so that Reaganism has, at least for the present, a less harsh and strident tone than the more-or-less open class warfare across the water. It also presides over a much stronger economy. But Reagan and Thatcher are fundamentally similar in their economic and social philosophies, the interrelationship of which is perhaps most clearly exposed in George Gilder's best-selling manifesto, *Wealth and Poverty.*[1]

It is, however, of doubtful value to distinguish between a confused economic and a coherent social strategy. This is not only because, in its view of women, the family, and teenage sexuality, among other questions, the social philosophy is no less a "rag-bag of hopes, hunches, prejudices and dogmas" than the economic. It is also because the social strategy, which aims at producing a climate favorable to capitalist enterprise, is an inseparable part of the Reagan-Thatcher political-economic program. Within an overall assault on public expenditures not for the military and police, the attack on the social services in both countries has specific economic and social (or ideological) purposes, and it is to these that we must now turn.

THE ROLE OF THE SOCIAL SERVICES

The cuts in the social services are one part of a strategy of rationalizing and restructuring social welfare to bring it into line with the "needs" of the economy, in particular the needs for capital accumulation and a labor market undistorted by either disincentives to work or trade union monopolies. This requires spending reductions, privatization, and a reorganization of incentives. It rests upon an implicit view of the political economy of the social services with roots in nineteenth-century economic liberalism but with some newer twists.

Economists have long been haunted by the Malthusian nightmare, a vision of a society in which charity or welfare so subverts the operation of the labor market that more and more people, acting rationally in their own interest, drop out of the workforce. The productive sector of the economy, constantly shrinking as a proportion of the whole, is in this scenario forced to support a growing burden of unproductive social expenses, thereby diverting the sources of further accumulation and impoverishing the entire society. Recent studies of unemployment benefits and the U.S. old-age insurance system have been informed by and reinforced that vision (Feldstein, 1977).

This line of argument extends beyond the recipients of social services to their providers, and indeed, to the whole public sector. Bacon and Eltis, whose book, *Britain's Economic Problem: Too Few Reducers* (1976), has had a large influence in Britain, argue that the growth of employment in the personal social services, local government, and education in the 1960s and early 1970s has significantly reduced the productive, "marketed sector" and increased the burden on it. They account for the rapid growth of social services as the result of attempts by governments to provide employment (reduce unemployment) cheaply, that is without much capital investment.[2]

Douglas Hague, one of Thatcher's policy advisers, makes a related point, which he modestly calls Hague's Law. It is based on a different sense of productivity (namely, output as a ratio of the amount of input required to produce it) and rests upon the lower rate of productivity growth in the services sector than the manufacturing, and hence of the public (heavily service-oriented) sector than the private. Hague argues that if public sector productivity rises (say 2 percent) less than the private, and the public sector's share of GNP stays constant (say 25 percent), and if pay comparability between the two sectors is maintained, the proportion of employees in the public sector will rise, and so will the taxes required to support them. That is, the private sector will use relatively fewer workers to produce its 75 percent of GNP, and the public sector relatively more to produce its 25 percent. Over 50 years, Hague calculates, the proportion of public employees and the tax rate will both double, to about 50 percent. (The paradoxical implication seems to be that the richer and more productive a society becomes, the fewer public services it can afford.) To counteract this effect, Hague argues, productivity should be more deliberately increased in the public sector (reducing the teacher-student ratio is one way the British government is attempting this in education). Equally important, public expenditure cuts should be a normal, ongoing process, not a temporary response to adverse economic circumstances (Hague, 1980). One aim of the Reagan-Thatcher strategy, which has achieved considerable success, may be said to be the bringing about of a permanent rhetorical shift in public debate, such that budget-cutting becomes the taken-for-granted norm. The issue becomes not whether to cut, but what, how, and how much.

Another implication of "Hague's Law" is that services provided by social workers are potentially as great a threat to capital accumulation as transfer payments, despite the much greater size of the latter. They represent a real command over resources by the government in a way that transfer payments do not (Gough, 1979: 108-117), although admittedly, of minor significance compared with that represented by the military. Until the recent reversal, however, military spending had been declining relative to GNP, as social service spending had been rising.

Finally, Hague's Law implies that pay comparability between public sector workers is untenable in the long run if one is intent on reducing the relative size of the public sector. One of Thatcher's most intensely fought battles has been over civil service pay claims, though she has failed to end their upward movement thus far. Even if she did, the resulting decline in the qualifications of workers attracted to government employ might have its own adverse effects on the productivity of public services.

In response to these and other pressures perceived as generating the inexorable growth of social services and their adverse effects on the labor market and investment, the Reagan and Thatcher policies call not only for the reduction of public spending but also for its restructuring so that social welfare grows more slowly than government expenditure as a whole — a reversal of the pattern of the previous three decades.

Similar considerations also dictate the policy of privatization, both as a means of achieving cutbacks and as valuable in its own right. Privatization takes the form of higher user charges (for example, in Britain, increased rents for public housing, higher prices for school meals, home helps, and National Health Service prescriptions). It also refers to the process of encouraging the growth of the private sector at the expense of the public, as in the Conservative policy of selling off National Health Service hospitals while relaxing controls and providing tax subsidies to stimulate the building of private hospitals. A tendency to reprivatization of the British postwar welfare state, to increasing selectivity and proliferating means tests was already apparent in the 1950s and 1960s (Briggs, 1961; Marwick, 1968), but it has been greatly accelerated under Thatcher.

The American welfare state, by contrast, has never had such a universalistic character as the British. Nevertheless, one aim of the present cutbacks is to force a finer discrimination of the "truly needy" and to encourage those who can provide for themselves through the market, or through informal or voluntary charity, to do so. Privatization reinforces the private sector both by reducing its tax burden and by increasing the dependence of workers and their families on the wages they earn. The latter point is important not only because of its social control aspect (as a buttress to managerial authority), but also because wages reflect the value of individual units of labor power (i.e., workers) much more closely than does the social wage. Despite the tendency of some welfare state programs to provide the highest benefits for those who need them least (as seen most clearly in earnings-related public pension systems, or in government support of higher education), social welfare subverts the operation of a wage system that disregards individual or family needs. Conservatives like Milton Friedman are, then, able both to denounce social security (in particular, old-age insurance) in the United States as the "poor man's welfare payment to the middle class," and to advocate stronger reliance on market mechanisms that disregard differences of individual circumstance altogether (Cohen and Friedman, 1972; Friedman, 1972).

Privatization and increased dependence on the wage system may also be seen as reinforcing the family. Critics of left and right have frequently denounced social service professions for usurping family functions and authority. In the same way, transfer payments are seen as substituting for the male breadwinner, making him redundant to family survival. Women are, of course, the major providers and consumers of the social services and, insofar as privatization results in a reduction of the total social service effort, public and voluntary, it will slow or reverse the process by which women's traditional functions of nurturing the sick, elderly, children, and so on have been professionalized and removed from the home.

A central part of the Reagan-Thatcher strategy is the governmental manipulation of individual behavior through modifying incentives to work and save. Regressive tax cuts to stimulate investment and cuts in unemployment benefits to stimulate work are the most obvious examples. But if home helps or meals-on-wheels are more expensive or less attainable, adult children will have an incentive to work harder to provide for their elderly parents. Those who want to be independent of their children in old age will have an incentive to save more, thereby contributing to net capital formation. The benefits of economic insecurity thus extend into every aspect of social and economic life!

THE PROBLEMS

The Reagan-Thatcher strategy undoubtedly has both intellectual and popular appeal, especially in the context of the failures of the welfare state and Keynesian economics to achieve their own objectives. But the impression of coherence and rationality tends to fade in the light of problems of implementing the strategy. Ian Gough describes elsewhere in this volume many of the dilemmas Thatcher faces in bringing her plans to fruition. We will not repeat the general points he makes there about the threats to capital accumulation, social reproduction, and legitimation and law and order implicit in the current policies. Instead, we will focus on some specific ways in which Reagan's and Thatcher's policies with regard to the social services are likely either to fail of successful implementation or to undermine their own goals.

One difficulty Thatcher and Reagan face is that, in their reliance upon local governments to determine the specific form of the cuts and the nature of the remaining services, they may not be able to monitor, much less control, the effects of their policies.[3] Furthermore, there is the question of whether or not local governments can be restrained from substituting their own taxing and spending for that of the national government. The emergence of such a "displacement effect" would disrupt national economic strategy. Though there is evidence of these problems in both countries, they are likely to be more severe in the United States than in Britain.

Whitehall has much more extensive control over local finance than does Washington. The rate support grant system provides the central government with a powerful lever over local authority current spending and it has been recently supplemented, as we have noted, by the institution of cash limits. Furthermore, central government has au-

thority to sanction local government borrowing for capital improvements (Judge, 1978: 31-66). Finally local authorities, who depend on
central government for the bulk of their financing in any case,
are compelled to rely exclusively on the property tax (or "rates") as
their only source of independent revenues (Newton, 1980). Even so,
there is evidence that local authorities are resisting these controls in
the face of pressure to maintain or increase services. Some authorities
have boosted their property taxes; others have begun to lobby the
Department of Environment as if they were interest groups instead of
administrative units of the British state.

Conditions in the United States suggest that the displacement of
central government spending to subnational governments will be
much more pronounced than in Britain. The United States government bears no direct responsibility for state and local finance. The
system of grants-in-aid, which proliferated during the postwar period,
involves Washington in funding a wide range of categorical programs
administered by the states and localities under a variety of more or
less carefully prescribed conditions (Reagan and Sanzone, 1981).
Many of these programs have developed powerful constituencies
made up of coalitions of clients, service providers, and local and
national politicians. The runaway grant system is rightly seen by
conservatives as a key element in the expansion of social spending in
the United States. Reagan's block grant proposal is supposed to break
this dynamic by ending program-by-program funding decisions. The
states, for their part, are promised grants with fewer strings attached
and, consequently, lower administrative costs to partly offset their
smaller size. The strategy assumes that state governments are more
willing to effect service economies than are federal bureaucrats,
members of Congress, or local governments. Already, this plan is in
doubt.

First, as we pointed out, Congress mangled the President's block
grant proposals. Second, the nation's mayors have been sounding the
alarm, fearing that monies funneled through state legislatures will
have their own strings, political and otherwise. Finally, even the
governors have expressed concern that the promised administrative
efficiency savings may not materialize and that they will be forced to
make up the difference from state revenues. There are certainly
grounds for claiming that Reagan achieved his own spectacular
budget cuts at the expense of the states. His proposal to appropriate
$86.4 billion in grants to the states is down 13.5 percent from Carter's
last budget and fully 66 percent of his proposed savings come from the
grants category (Congressional Quarterly Weekly Report, April 25,
1981: 709).

Many of the cuts in social services will not clearly save public money in the long run. There will be instead another kind of displacement effect. Thus, cuts in the child nutrition programs will be severe, despite the evidence of their importance for improved pregnancy outcome and child health, learning, and development. These reductions may produce long-term economic costs by undermining the reproduction of efficient labor power (with possible adverse effects on international competitiveness). In the shorter run, they are likely to increase the costs of other social services — those of health care, as well as rehabilitative, remedial, and even institutional services for those with developmental delays, or learning or behavior problems. As several studies show, the WIC program (which provides supplementary food, nutrition, education, and health services for pregnant and lactating women, for infants, and for young children) has a powerful effect on the incidence of low birth-weight infants, anemia, and infant mortality. One study conducted at the Harvard School of Public Health found that each $1.00 spent on the prenatal component of the WIC program results in a savings of $3.00 in hospitalization costs, due to the reduced number of low birth-weight infants needing extended hospital care (Massachusetts Department of Public Health, 1981; USDA, 1981). Such evidence persuaded the Senate to exempt the WIC program from the first round of cuts. Similarly, the British directors of social services warned the Conservative government last year that further cuts would have "very serious consequences and indeed, will endanger and foreshorten lives and will put considerable increased pressure on the health services" (New Society, September 18, 1980: 564).

It is not even clear that such assaults on the living standards and quality of life of the poor will increase the incentive to work. The working poor, especially working women with dependent children, are among the hardest hit. Loss of day care, work expense allowances, and other services, are likely to force many women, and encourage many others, to relinquish their low wage or marginal employment and drop out of the workforce altogether. Many are likely to stop working in order to requalify for benefits and services on which they depend. As Lester Thurow puts it, "Suppose you are one of the working poor and have a sick child. One choice is to work harder — perhaps by taking a second job — in order to pay the necessary medical bills. Another is to quit to make yourself eligible for Medicaid. To pose the choices is to give the likely answer" (Thurow, 1981: 8). This "poverty trap" will also raise the cost to the poor of getting a job or returning to work, and is likely to prove a powerful disincentive to labor force participation (Field, 1980).

PROSPECTS FOR OPPOSITION

The prospects for effective opposition to Reagan's and Thatcher's social service policies are not bright — at present. As Hugh Heclo observes, postwar prosperity has undermined the sense of social insecurity as a shared danger. "Major dislocations such as depression and war had struck every social group; minor fluctuations in a period of overall growth touched only certain sections" (Heclo, 1981: 395). The social distance between most of the working class and the poor has grown, reinforcing individualist and antiwelfare attitudes. The perception of a conflict between individual interest and state welfare has, as Peter Taylor-Gooby (1981) argues, a material basis, and is reinforced by individual experience of social services as often demeaning, intrusive, and inadequate.

At the same time, those fears of the neoconservatives, public choice theorists, and others about a "revolution of rising entitlements," competition among parties to win votes by expanding social spending, uncontrollable budgetary obligations, and so forth, have proved largely groundless. Democrats and Republicans, Conservative and Labor parties have competed in lowering expectations and reducing the growth rates of taxes and social spending, in attacking inflation while tolerating high and growing unemployment (Heclo, 1981: 401-402).

The "opposition" parties in both countries are in disarray. Both are handicapped by their own records in office and neither is able to project a radically different approach. They are forced, therefore, to fight on the ground of their conservative opponents (see Miliband, 1980: 280). Labor is embroiled in an internal power struggle between left and right. Several leading figures in the party have broken off to launch a Social Democratic movement, but it is very unlikely to reject the Conservatives' efficiency and productivity-linked arguments against state spending. Potentially, the most effective resistance to Mrs. Thatcher comes from within her own party and from the business community, which is reeling from her policies. But these voices are calling for tax reductions, easy credit, and reflation, not a reinvigorated system of social protection. If anything, the American Democratic party is even less able to fight than Labor. Unable to deny the President his legislative victories, the party leadership appears content to let him have his day in court so that they will have an election issue if, as they expect, he fails.

Unions in both countries have been extraordinarily weak and ineffectual in the period from the mid-1970s. Workers involved in the services under attack show few signs of militancy. In Britain a national Fightback conference of health trade unionists produced an attendance of fifteen. Social workers remain weakened and demoralized by a long, unpopular, and unsuccessful strike in 1978-1979, one which served to persuade many of their dispensability. In the United States, social workers are much less organized into unions, while their professional organization has adopted a strategy of "realistic" legislative lobbying which has included implicit support for the lesser evil of Democratic budget cutting (Adams and Freeman, 1979; NASW News, May 1981).

On the other hand, the probable failure of Reagan and Thatcher to achieve their own goals, either with regard to the social services or the economy as a whole, will undermine their support and legitimacy. The process is already well under way in Britain. Translation of such unpopularity into effective political opposition, in a context where the opposition parties themselves introduced many of the policies in question when they were in power, is another matter.

But in abandoning the accommodationist and solidarist (or class collaborationist) strategies of Conservative and Labor governments over most of the last 100 years, Thatcher has embarked on a dangerous experiment. On the one hand, she has weakened the political basis of labor reformism and the trade union bureaucracy, as the deep and prolonged recession has undermined its economic basis. Union leaders are no longer able to offer much to the employers or to their rank and file, and Thatcher is not interested in their collaboration. They may be much less able to resist the pressure of a militant rank and file in the future.

On the other hand, Thatcher's disregard of the social control functions of the welfare state, its contribution to an ideology of social solidarity, to the legitimation of the capitalist economy and state, is likely to cost her dear. She has increased government spending on the unproductive forces of law and order, and in response to the recent wave of urban riots she has reversed herself on public spending for youth employment programs and other employment-related social welfare measures, to the extent of an additional billion dollars. The capacity of urban riots to elicit increased spending on social services and income transfers is well known in the United States, though perhaps temporarily forgotten (Piven and Cloward, 1971).

There is, then, reason to doubt that either Reagan or Thatcher has resolved the central problem of capitalist social policy, the development of a social welfare system that contains the discontent of the working class and the poor as cheaply as possible, while at the same time being optimally adapted to the needs of the economy.

NOTES

1. For a useful and entertaining discussion of the social and political assumptions of Gilder's economics, see the review of his book by Michael Walzer (1981).

2. For a critique of Bacon and Eltis, and for a discussion of the "return flow" by which public sector workers *contribute* to the private sector, see Gough (1979: 106-22).

3. Thus, Patrick Jenkin, British social services secretary, has been criticized in this respect by the House of Commons social services select committee. With both Conservative and Labor members, it unanimously challenged his lack of "strategic policy-making" and the "failure to examine the impact of changes in expenditure levels." Even the Department of Health and Social Services has depended for information on such sources as the surveys conducted by the association of local directors of social services among its members about how each authority had responded to the cuts (*New Society,* September 18, 1980: 547).

REFERENCES

ADAMS, P. and G. FREEMAN (1979) "On the political character of social service work." Social Service Review 53 (December): 560-572.

BACON, R. and W. ELTIS (1976) Britain's Economic Problem: Too Few Producers. New York: St. Martin's.

BENDIX, R. (1956) Work and Authority in Industry. New York: John Wiley.

BRIGGS, A. (1961) "The welfare state in historical perspective." Europäisches Archiv für Soziologie 2: 221-258.

COHEN, W. and M. FRIEDMAN (1972) Social Security: Universal or Selective? Washington, DC: American Enterprise Institute.

DERTHICK, M. (1975) Uncontrollable Spending for Social Service Grants. Washington, DC: Brookings.

FELDSTEIN, M. (1977) "Social insurance," in Colin D. Campbell (ed.) Income Redistribution. Washington, DC: American Enterprise Institute.

FIELD, F. (1980) "Mrs. Thatcher's poverty trap." London Observer June 8.

FRIEDMAN, M. (1972) "Social security: the poor man's welfare payment to the middle class." Washington Monthly 4, 3: 11-16.

GOUGH, I. (1979) The Political Economy of the Welfare State. London: Macmillan.

Great Britain (1980) The Government's Expenditure Plans 1980-81 to 1983-84 (Cmnd. 7841). London: HMSO.

——— (1976) Cash Limits on Public Expenditure (Cmnd. 6440). London: HMSO.

HAGUE, D. (1980) "Why the public expenditure cuts must go on." London Times (June 20).

HECLO, H. (1981) "Toward a new welfare state?" in Peter Flora and Arnold J. Heidenheimer (eds.) The Development of Welfare States in Europe and America. New Brunswick, NJ: Transaction Books.

JUDGE, K. (1980) "Is there a 'crisis' in the welfare state?" Presented at the University of Texas at Austin.

——— (1978) Rationing Social Services. London: Heinemann.

MARWICK, A. (1968) Britain in the Century of Total War. London: Bodley Head.

Massachusetts Department of Public Health (1981) "Effect of WIC on neonatal mortality and other outcomes of pregnancy: preliminary results." Boston.

MILIBAND, R. (1980) "Class war conservatism." New Society 52 (June): 278-280.

NEWTON, K. (1980) "Central government grants, territorial justice and local government," in Douglas E. Ashford (ed.) Financing Urban Government in the Welfare State. London: Croom Helm.

PIVEN, F. F. and R. A. CLOWARD (1971) Regulating the Poor. New York: Random House.

POLANYI, K. (1944) The Great Transformation. New York: Holt, Rinehart & Winston.

Political Quarterly (1981) 52 (January-March).

REAGAN, M. D. and J. G. SANZONE (1981) The New Federalism. 2nd ed. New York: Oxford University Press.

RICHAN, W. C. (1981) Social Service Politics in the United States and Britain. Philadelphia: Temple University Press.

TAYLOR-GOOBY, P. (1981) "The new right and social policy." Critical Social Policy 1 (Summer): 18-31.

THUROW, L. (1981) "How to wreck the economy." New York Review of Books 28 (May 14): 3-8.

TOWNSEND, P. (1979) Poverty in the United Kingdom: A Survey of Household Resources and Standards of Living. Berkeley: University of California Press.

United States Department of Agriculture. Office of Policy, Planning and Evaluation, Food and Nutrition Service (1981) "Evaluation of the effectiveness of WIC." Washington, DC. (mimeo)

WALZER, M. (1981) "Life with father." New York Review of Books 28 (April 2): 3-4.

5

Determinants of State Housing Policies: West Germany and the United States

PETER MARCUSE

□ WEST GERMANY AND THE UNITED STATES afford a comparison that highlights the real role of the state in housing. The size of the state sector is not only much greater in Germany[1] than in the United States (tax revenues are 38.2 percent of GNP compared to 30.3 percent in the United States [U.S. Department of Commerce, 1980: 906]), but state activity is also more extensive, more explicit, longer-term, and more widely accepted. Therefore, the causes and consequences of state actions are also more visible in Germany — or at least visible in a light very different from that of the United States. The comparison with the United States, where governmental intervention seems to be less extensive than in any other developed private market economy, is particularly illuminating. The thesis here will be that, despite differences in magnitude and form, the role of the state and the determinants of its actions are in fact the same in both countries. A general formulation of that role is first put forward as a hypothesis, and then tested against the varying experience of the two countries.

DETERMINANTS OF HOUSING POLICY

Any comparative study must have an organizing framework of hypotheses as to underlying principles. Otherwise, the comparison becomes simply a painting of individual cases to be pinned up against a wall side by side, from which a reader may or may not be able to draw some conclusions, unaided by the author (see Walton and Masotti, 1976). There have in fact been a number of recent studies that do precisely this. Others, substantially more analytic, focus on one or another explanatory factor and look only at the outcomes that seem relevant to understanding (or discounting the importance of) that factor.[2] The lack of a general theory of housing is partially responsible for the difficulties. This chapter should be seen only as an attempt to move the common exploration forward one step, primarily on the level of theory rather than description.[3]

The broad explanation of housing policies that underlies the comparative account herein roots state policy in basic conflicts.[4] Housing policies[5] are determined by the outcomes of conflicts among broad sectors of society. Many explanations acknowledge conflict, but see it as a constraint on an underlying effort by government to solve the housing problem (the Myth-of-the-Benevolent-State view). Other explanations acknowledge conflict, but portray the situation as one in which the private sector does its thing, builds, operates, tears down, relocates as it wishes, either independently of government (the neoconservative Myth-of-the-Autonomous-Private-Sector view) or fully in control of government (the exaggerated structuralist neo-Marxist view), and conflicts at most temper its effectiveness. The argument put forward here is that conflicts are central to events, not marginal to them.

Nor are the conflicts that determine housing policies confined to the housing sector — quite to the contrary, housing is only one stake in broader conflicts, and often one of the smaller stakes.

The tendency of much contemporary writing on housing to deal with it in isolation, or to see it as the focal point of a defined set of policies designed to deal with various problems associated with it, is a major weakness. It is a weakness that is ideologically loaded, for it conceals the fundamental mechanisms of society from view, revealing a bit here and a bit there, but distracting attention from any attempt to view these mechanisms as a whole.

Specifically, conflicts about housing in private market societies are a component of broader conflicts between classes, essentially defined by their positions within the relations of production in that society. Both private and state actions having to do with housing are in the first instance determined by the particular status of these relations, those occupying particular positions within them, those structures (political, economic, social, ideological) produced by them. Whether housing is publicly or privately provided; publicly, privately, or communally owned; is very much dependent on these broad relations of society. If it is private, how much an individual can afford to pay for it is determined by the position the individual occupies in the hierarchy of jobs and the wealth the individual has accumulated or inherited under the legal and economic arrangements dominant in that society. Where housing is located is determined by where the productive facilities and their ancillary services are located. The quality of housing for various groups will be determined by how those groups share in the product of that society. The standards generated for the provision of housing and of other necessities of life are the result of conflicts on a more general social plane. The costs of housing, who provides it, what provisions for security of tenure exist, what rights accompany it, who profits from it, what services it offers, all depend on the particular way the major forces in that society have adjusted their particular differences, what accommodations have been made in terms of wages, prices, the size of the public sector, the political rights of citizens, the extent of participation in decision-making — all accommodations (or ongoing conflicts) worked out on a much broader terrain than just housing.

TYPES OF CONFLICT

If conflicts determine housing policies, what are these conflicts about? The profit to be made from the provision of housing, and the amenities to be enjoyed by its use, of course spring to mind. But this is only one of the stakes. A general formulation of the subjects of conflict might read:

(1) profitability in the economy as a whole (variously referred to as the long-term accumulation of capital, or national economic growth);

(2) political power (stability of the political system, hegemony, legitimacy of the established order);

(3) housing sector profitability (profit from the construction, financing, ownership, and management of housing); and

(4) residential use (the individual and collective consumption of housing[6]).

These are of course artificial categories, and their separation for purposes of discussion and analysis might give rise to the impression that they can be separated in reality, that they can be pursued independently and without critically interacting with each other. That would be unfortunate, for the essence of the argument made here is that they are integral. The separation is intended only as an expedient to come to grips with the underlying forces, to show the unity of events, forces, stakes in the conflicts around housing policies.

The four areas of conflict may be more precisely defined, for housing, as revolving around:

(1) Efforts, in the economic sphere, to increase the profits of business activity in general, with basic housing a necessity for the essential work force, as well as a possibility for social control, but the cost of housing also exerts a to some degree undesirable upward pressure on wages, as opposed to efforts to improve the real standard of living of workers;[7]

(2) efforts to use housing to support the political status quo and avoid political unrest growing out of dissatisfaction with housing, as opposed to demands for adequate housing as a political/civil right;

(3) conflicts over the division of profits among various groups directly involved in the production of housing, reflected in the prices of housing and its components; and

(4) conflicts over the uses and the amenities of housing, involving for instance, location issues, standards, social patterns, and provision of collective facilities and services.

These conflicts, we hypothesize, determine events in the housing sector. They determine both private and public actions. Whether the resolution of any given conflict finds its expression in a private market arrangement or in a governmental action depends very much on who won and which approach best serves to implement that victory, and varies significantly from case to case, country to country.

Yet whether the resolution is carried out through state actions or not is not simply and exclusively determined by the parties to a given conflict. The state itself (not in disembodied juridical form, but personified in those that directly lead the state and carry out its mandates: political leaders, appointed officials, bureaucrats, judges, and so on) has a history and a set of determinants governing its actions that must be taken into account. The representatives of the state are not simply pawns in the games of others, but players themselves. The interests of elected officials in remaining in office, civil servants in keeping their jobs or increasing their power, informal dependencies on political contributions or social status, and the particular motivations (and ideals) of individuals within the state apparatus, all come into play.

STATE ROLE

Two types of activities lend themselves particularly well to execution by the state: the rationalization of atomic, separate, private activities, and the mediation among equals within dominant forces. Rationalization is the kind of activity assigned to the state by welfare economics: dealing with externalities, providing for public goods. City plans, street layouts, mass transit, building standards, even the pooling of mortgage risks or the assembly of land, may be functions that the state, because of its legal powers and supraindividual character, can carry out better than the market.[8] Government can, and does frequently, mediate conflicts among groups within the housing industry, financial interests against builders on interest rates, for instance, or craft unions against builders on construction standards, or competing pressures for loft space between small manufacturers and residential developers. Thus, the role of government is decisively, but not exclusively, determined by private interests and private decisions.

The argument as to the role of the state in housing can be summarized — subject to and to the extent consistent with the resolution of the conflicts described above — as:

(1) the service, in a subordinate capacity, of the forces that control the state, by
    (a) supporting profit making and accumulation in the economic sphere,
    (b) supporting the continued control of the dominant powers in the political sphere, and
    (c) providing and allocating residential benefits;

(2) the quasi-independent service of such forces, by
    (a) promoting rationalization of private activities, and
    (b) mediating conflicts; and

(3) the protection of the self-interest of the components of the state apparatus.

Specific state actions affecting housing arise from the tension among and within these basic roles. The state's service role is likely to be a largely passive reflection of these tensions; its rationalization, mediation, and self-protection activities will initially be more independent.

The ultimate determination of housing policies results from the interplay of these activities, but that interplay does not have random results. The order of priority is hypothesized to be as follows:

— First, the maximization of profits in the economy overall — the guarantee of accumulation — will be the single most continuous and generally most important determinant.

— Second, when housing becomes a critical political issue, the need for stability, for legitimation, will take priority.

— Third, the relative distribution of profits to be made as among those within and outside the housing field, and the arrangements for the residential use and enjoyment of housing as a consumption good, will in most cases, and subject to the above overriding priorities, determine short-term housing policy.

To reformulate the same point: Conflicts around long-term accumulation and political stability are the underlying determinants of housing policy; within the broad parameters allowed by their resolution, the interests of housing producers, housing consumers, and the state apparatus will determine the specific shape a given policy will take.

The following discussion examines to what extent these determinants of housing policy have operated in the very different West German and U.S. context in the postwar years.

HOUSING POLICIES IN WEST GERMANY

The basic thrust of housing policies in Germany has been very much like that in the United States: to permit maximum scope and support to the private housing industry (property owners, builders, suppliers, finance, management), with governmental regulation and subsidy limited to supporting that industry except for those situations in which other overriding priorities exist. The normal situation is the production, financing, and management of housing for private profit, which means in response to economically effective demand in the private marketplace.

Other overriding priorities, however, have been much more prominent in Germany since the war than in the United States. The actual policies adopted have therefore differed widely in the two countries. In West Germany, four such overriding priorities have been:

(1) the critical (in a physical, economic, and political sense) shortage of housing immediately following the war;

(2) the need for resources to rebuild industry and restore production in the economy as a whole, requiring a limitation of resources committed to housing;

(3) the ongoing stability and prosperity of the economy as a whole, requiring adjustments to investment in housing as the business cycle and trade relations might dictate, with a particular concern for Germany's competitive international position; and

(4) the concern to maintain political stability and avoid breaching basic social harmony because of the housing issue.

These are not, of course, concerns independent of each other: In the immediate postwar period, for instance, meeting shelter needs and establishing political confidence in the new government were hardly separable.

DEVELOPMENT OF GERMAN POLICY

The big picture in the Federal Republic might be painted as follows: In the first postwar years, the overriding priorities were of such importance as to cause government to impose severe restraints on the housing industry. The need to build housing quickly, and the need for resources in other sectors also, suggested a major governmental role; political/ideological considerations, including relations with East Germany and the campaign against communism, suggested that the visibility of the state as such be minimized; the recreation of a private market in housing remained the long-term goal. The policy of major subsidies, but channeled through nongovernmental nonprofit institutions, provided the answer. As the housing shortage became less critical in the late 1950s, political stability under Adenauer more secure, and competitive demands for resources more effectively expressed through the market, the underlying thrust toward the private sector emerged more strongly, both in the expansion of unsubsidized housing and in the increase of the private role within the subsidized sector, e.g., in financing. Increasing attention was focused on higher-income households and on home ownership as opposed to rentals, both of which permitted a lessening of the government's direct role. This shift back toward the "normal" private market can be dated as commencing in a major way in 1956 with the Second Housing Construction Act, and reaching a high point in the mid-1960s, with the end of rent control and the introduction of the Second Subsidy System. The increasing role of tax benefits as opposed to direct subsidies reflects the same developments. So does the largely unregulated rebuilding of the central areas of many cities, converting residential uses from lower- to higher-income and residential uses in general to commercial.

Political unrest shook most of Europe, as well as the United States, in the mid-1960s, culminating in 1968, and led to the election of a Social Democratic government in Germany. Protest against the effects of unbridled privatization waxed strong and influenced housing policies. Delays in the ending of rent controls, increased levels of expenditure (although increasingly channeled through housing allowances), avoidance of some inner-city housing demolitions, and special housing investment programs, were partly the result. Some of these results were short-lived; others were more permanent, as for instance the institutionalization, with provisions for citizen participation, of urban renewal and modernization, and the adoption of rent stabilization for private housing.

With the ending of large-scale unrest, and the economic downturn of 1974, this deviation from the central thrust of policy diminished. The more "normal" situation reasserted itself in Germany (as it did in the United States in 1972); a renewed focus on the role of the private sector, a reduced level of government subsidy, increased play for private market forces in housing (including, on the occupant side, greater attention to middle- and upper-income households and to home ownership), and renewed (if more carefully planned) adjustment of land uses in the inner-city areas to the changing demands of the local market. But the experience of the late 1960s was not forgotten; through carefully tuned policies of citizen participation, of income-based housing subsidies, of rehabilitation purporting to balance the interests of private investment and low-income tenants, of refined tax policies, the government hoped to avoid the direct confrontation and protest that had proved so troublesome in the earlier period.

But opposition to current policies is growing; fiscal cutbacks may heighten dissent; confrontations are on the increase. Whether the approach of fine-tuned compromise will work as intended or not remains to be seen.

To summarize: Postwar housing policies in West Germany might be explained through the following history:

(1) phase I — a postwar emergency and reconstruction, in which the private housing sector was fully subordinated to broader social and economic interests, with heavy governmental involvement (1949 to about 1953);

(2) a transitional stage, in which governmental controls were re-
    laxed as housing users' most urgent needs were met and the
    business community could more easily stand on its own feet
    (about 1953 to about 1960);

(3) phase II — a period of large-scale reprivatization of the housing
    sector, giving the housing industry encouragement to expand its
    role in a "healthy" private market (about 1960 to about 1967);

(4) resistance by housing users to these changes, direct conflicts
    involving labor and threatening both to the business community
    and to governmental leaders, resulting in a change in that lead-
    ership and a restraining of certain private activities in the hous-
    ing sector (about 1967 to about 1973);

(5) phase III — compromise and the attempt to institutionalize the
    results of that compromise in a set of comprehensive and fine-
    tuned housing and community development policies (about
    1973 to about 1979); and

(6) a new period of conflict, in which diverse users of housing
    manifest their dissatisfaction with the existing situation, with
    widespread support in many localities but with a heavy array of
    national forces against them (about 1979 to the present).

SPECIFIC POLICY CONTEXT

To present the story in more detail, then:[9] The housing supply
available in West Germany in 1950 was at a catastrophic level. Only
an estimated 10.1 million units were available for at least 16.7 million
households. And the quality of many of those that did exist was sorely
lacking; for instance, less than 20 percent had either a shower or a
bath indoors for the exclusive use of the occupant.

Both the drastically reduced level of housing construction during
the Depression and under Hitler, as well as the devastation of the war,
had contributed to the shortage. After the virtual cessation of housing
production during World War I and the inflation that followed it, the
1920s had seen a dramatic upsurge in housing construction and in the
level of governmental support for the improvement of housing conditions. Housing construction never regained this pre-Depression volume under the Nazis. Furthermore, the level of public investment in
housing as a proportion of total housing investment went down sharply in Germany (unlike other Western countries) in the 1930s. From
1924 to 1931, the public share never dropped more than a percent under
40 percent; between 1932 and 1938, it was never above 25 percent;
from 1939 to 1946, it virtually disappeared.

The devastation of the war, added to the cumulative shortfall in units not built during the Depression nor under the Nazi government, thus created a catastrophic housing situation. Furthermore, the resources needed for housing construction were severely limited, and badly needed in many other sectors of the economy. The assistance offered under the Marshall Plan was a major factor in permitting the reconstruction that followed at such a rapid rate in the years after 1950. A third factor, immigration and forced resettlement first from East Germany and Eastern Europe, and then of foreign workers, also contributed heavily to the subsequent shortages; immigration, in fact, accounted for a substantial part (55 percent) of the population increase in West Germany after 1950.

PHASE I — POSTWAR PERIOD

The new constitution, adopted in 1949 under the watchful eyes of the Western allies, did not clearly delineate the future direction of policies affecting housing. An all-encompassing subordination of private property to the public welfare was expressed in Article 14, but a similarly comprehensive commitment to the protection of the rights of private property appeared in Article 15. A flexibility of action was thus created, to be used effectively thereafter.

The first two major actions of the new government to deal with the extreme housing shortage reflected that flexibility; first, the passage of the 1950 First Housing Construction Act (*Wohnungsbaugesetz*) and, second, the formal continuation of the rent freeze and housing controls. The 1950 Act was drafted in less than two months and then adopted almost unanimously by the legislative branch in record time. Under its provisions, direct governmental loans and grants were to be made at reduced rates of interest to specified types of housing developers, largely nonprofits, to build housing. Rent levels were set by statute and subsidies granted to individual projects at those levels needed to produce the prescribed rents.

The largest number of social housing units ever built in Germany in a single year, 317,000, was built pursuant to this legislation in 1952 (see Table 5.1). That number was 69 percent of total new construction, equivalent to 3.5 percent of the total standing stock of housing. For an order of magnitude: Social housing built in West Germany in 1952 equaled, proportionately, the total construction of *all* units in the *highest* production year in the postwar history of the United States (1972), and compares with an all-time high for subsidized housing construction in the United States of 0.6 percent of the total stock in 1971.

TABLE 5.1   Housing Construction in West Germany, 1950-1979

(1) Year	(2) Total number of units completed	(3) Single-family units as percentage of (2)[a]	(4) Social housing as percentage of (2)[b]	(5) Lower-income social housing as percentage of (4)[c]
1950	371,924	35.2	68.6	100
1951	425,405	35.2	69.5	100
1952	460,848	33.4	68.9	100
1953	539,683	32.5	56.4	100
1954	571,542	34.2	54.2	100
1955	568,403	36.3	50.8	100
1956	591,082	37.2	51.7	100
1957	559,641	39.6	52.4	100
1958	520,495	40.2	51.7	100
1959	588,704	39.2	51.2	100
1960	574,402	40.3	45.8	100
1961	565,761	42.8	42.8	100
1962	573,375	42.4	42.3	100
1963	569,610	43.3	40.2	100
1964	623,847	42.6	39.8	100
1965	591,916	44.1	38.6	100
1966	604,799	42.9	33.6	88.1
1967	572,301	41.9	33.7	86.0
1968	519,854	41.1	34.2	73.3
1969	499,696	40.7	36.7	79.4
1970	478,050	41.0	28.7	78.2
1971	554,987	40.4	26.8	81.8
1972	660,636	37.6	23.2	69.4
1973	714,226	36.8	23.7	60.8
1974	604,387	38.0	24.5	64.3
1975	436,829	44.7	29.0	58.5
1976	392,380	53.0	32.6	46.9
1977	409,012	55.4	34.1	50.9
1978	368,145	64.7	24.5	40.9
1979	357,766	66.0	26.2	45.1

SOURCE: Institut Wohnen und Umwelt, *Grunddaten*, Darmstadt, 1980.
[a]Owner-occupied units in 1- and 2-family houses; estimated, for 1950-1962.
[b]First and Second Subsidy System.
[c]First Subsidy System as percentage of all social housing (authorizations).

The continuance of the wartime (actually prewar) rent freeze was part of the effort to avoid an inflationary and speculative reaction to the scarcity of housing, at a time when the economy could not afford such a diversion of resources to nonproductive use. The passage of the Land Acquisition Act of 1953 (*Baulandbeschaffungsgesetz*), which made statutory the limitation on the price to be paid for land required for the construction of housing, was part of the same effort. In both cases, values as of 1936, the date on which earlier price freezes were placed in effect, were the basis for the current price (or rent) to be paid, with only limited adjustments. It was widely recognized that the resultant figures were far below current market value, and that eventual retreat from these strict rules would be required; but under the emergency conditions existing in the earlier 1950s, the limitation on profits was justified by the government. The interests of the private housing sector, including landowners, had to be subordinated to overriding priorities: political — housing for those most desperately in need (usually lower-income households); and economic — housing for those most desperately needed (usually factory workers).

The adoption of the 1953 amendments to the First Housing Construction Act, and even more, the passage of the 1956 Second Housing Construction Act (II *Wohnungsbaugesetz*), signaled the gradual relaxation of the earlier emergency measures and a major shift back toward the private market. The 1953 Act lifted rent restrictions imposed earlier on tax-assisted (but otherwise unsubsidized) housing. The 1956 Act proclaimed the achievement of home ownership by "broad sections of the people" (the statutory phrase) a major goal of public policy. Subsidies were reduced, private capital was more heavily relied on, incentives to private involvement were increased, and restrictions on eligibility for tax benefits eased. The 1955 Federal Rent Act provided for the gradual increase of rent ceilings throughout the country.

PHASE II — REPRIVATIZATION

In 1960, the end of the postwar emergency was formally proclaimed with the passage of legislation appropriately called the Act for the Dismantling of the Housing Emergency System (*Gesetz über den Abbau der Wohnungszwangwirtschaft*). The Dismantling Act signaled the completion of the transition to phase II of postwar West German policies affecting housing. It was explicitly intended to reintegrate the housing sector into the private market economy. It adopted as a matter of policy the termination of rent controls and tenant protection legislation, although the implementation of the termination was temporarily postponed in anticipation of a gradual phasing out.

The Federal Building Act, also passed in 1960, lifted the earlier freeze on the price to be paid by public bodies for land, but provided for only very limited guidance by government of ongoing extensive private city development activities. It codified governmental planning and building rights, land-use controls, and powers of eminent domain, seeking to permit the free market to function again in a predictable environment. Substantial amendments to the private market orientation were adopted in 1971, 1975, and 1978. These responded to the pressures both of grass-roots citizens' protests in the late 1960s and changing economic needs requiring adaptations in urban form. In 1960, however, the trend to privatization of housing, planning, and land use held uncontradicted sway.

Other measures in this period of privatization in the early sixties had a similar orientation. The Federal Rent Act of 1960 (II *Bundesmietergesetz)* permitted the trend toward rising rents in the private sector to continue. Even within the publicly subsidized social housing sector, rents were increased. They began to be based on actual costs of operation and not, as previously, on a statutorily determined rent ceiling. This change was foreshadowed in the legislation of 1960, and codified in the Act for the Regulation of Housing in 1965 *(Wohnungsbindungsgesetz).*

The Housing Allowance Program (Wohngeld) was also expanded significantly in 1965. The philosophy of the program implies the strict limitation of subsidy to specific income groups based on need, as opposed to an ideologically more far-reaching general obligation of government to provide housing for all groups as a matter of right. Further, the shift from supply- to demand-side subsidies enabled the housing industry to act as the private market prompted, rather than under immediate governmental direction. Housing allowances arguably are principally a government subsidy to private landlords, permitting rent increases not otherwise possible. The introduction of allowances in 1963 occurred simultaneously with the ending of rent controls, lending some support to this interpretation; early research in Düsseldorf, however, did not reveal any rent-increasing effect of allowances.[10]

A shift in the form of the basic subsidy for social housing, from capital loans and grants directly from government funds to subsidies for the payment of interest and principal on private funds, can be interpreted as part of the same revived private thrust. It was first authorized by the Second Housing Construction Act of 1956. By 1967, the annual contribution made by such current subsidies for private loans exceeded the annual benefit of permanent government loans and grants. The Second Subsidy System, adopted in 1967 to extend the benefits of social housing to higher-income households and at the

TABLE 5.2   Distribution of Governmental Housing Benefits, by Income,
             West Germany

(1) Income Group	(2) Income Group as Percentage of all Households	(3) Percentage of Housing Benefits Received
Under 800 DM	22.0	12.3
800-1200	25.3	18.6
1200-1600	20.5	20.4
1600-2000	14.1	16.7
2000-2500	9.0	13.1
2500-3000	3.8	6.9
3000 plus	5.2	12.1
Total	100.0*	100.0*

SOURCE: Rudi Ulbrich, Wohnungspolitische Ziele und personelle Verteilungswirkungen wohnungspolitischer Transfers in der BRD, Augsburg, 1980, as set forth in IWU, *Grunddaten*, supra.

*rounded

same time reduce the level of per unit subsidy, was financed entirely by such current contributions to meet the costs of borrowing privately. It accounted for a steadily increasing share of all social housing (see Table 5.1). Thus, private capital, now more significantly available, played an ever-increasing role in the housing field, including subsidized housing. By providing for a staged reduction of current contributions for a given project, the concept of only temporary government involvement, and ultimately complete privatization, was further advanced.

Housing subsidies distributed through the tax system likewise served to reinforce, rather than change, private housing market patterns. Tax subsidies grew in importance throughout the 1960s, and continued to grow, shielded from public attention by their implicit nature, throughout the 1970s. It was not indeed until 1979 that the Federal Ministry of Housing put together figures showing the net distributional consequences of governmental housing policies, including tax benefits. Current figures are shown in Table 5.2. Tax benefits resulted from the increased use of the Savings Premium program providing tax deductions or premiums to those regularly saving for home purchase; the increased use of income tax benefits available to existing owner occupiers; and (although not until 1977) the extension of the Section 7b income tax deductions (a form of accelerated depreciation) to purchasers of existing as well as new housing. These all served to shift the benefits of government assistance in housing toward higher-income groups. The higher the income,

the more exclusively private are the household's housing arrangements. In effect, such government subsidies serve to buttress already
economically effective demand, stimulating and enlarging the scope
of the private housing market. In the best recent calculations, tax
benefits as a proportion of housing subsidies went up from 19.8 percent to 29.6 percent between 1965 and 1980, while social housing
subsidies went down from 53.5 percent to 32.1 percent.

The changing relationship between the public and private sectors
can also be seen in the area of housing finance. If one looks not at the
amount of the subsidy provided by government for housing (which,
net, has been increasing), but rather at the way it is provided, the
pattern of privatization is again clear. Rather than tax-produced funds
being loaned at no interest or below market interest rates, government
is now paying market interest to private financial institutions and
lenders who make the equivalent loans. From 1952, public capital
dropped from 43.1 percent of all capital for new housing to 29.1
percent in 1959, to 9.3 percent in 1968, and stood at 3.4 percent in 1978.
The change is similar to that in the United States from the Section
221(d)(3) to the Section 236 programs, and came about for much the
same reasons. Private lenders have increased their role and expanded
the range of safe and profitable lending through the change.

But these measures to recast the government's role in the housing
sector also produced problems for many households. Housing costs
rose faster than incomes for many. From a low rent-to-income ratio of
8.8 percent in 1955, the figure for a middle-income worker's family
rose just about 0.1 percent a year over the next 10 years, but then
climbed steeply, in 1965 by 1.0 percent. It rose to 10.9 percent in 1966,
to 12.1 percent in 1967, and to 13.5 percent in 1968. In 1966, political
pressure forced the postponement of the termination of rent control;
the recession of 1966-1967, with its attendant sharp reduction in the
level of housing construction, intensified the opposition to the reduction of governmental assistance in the housing sector.

The recession, escalating interest rates and concern about inflation, and the wave of militance, community organization, and protest
that swept Europe in 1967-1968, all ushered in a period of conflict and
flux that lasted for at least five years.

Electoral results reflected the change. In 1966, the Christian
Democratic party, which had ruled uninterruptedly since 1949, was
forced to share power with Social Democrats; in 1969, for the first
time since the war, the Social Democratic Party came to power,
sharing a majority with the Free Democrats in a coalition government
until 1972, then by itself. The shift in housing policies paralleled the
shift in government, but both reflected broader tensions in the society

as a whole. Serious, if not basic, conflicts about housing and planning policy took place in this period more than at any previous time since the war. The increasing number and militance of citizens groups *(Bürgerinitiativen)* were the most visible manifestations of that conflict. While many individually failed, the forces they represented had to be, and were, taken into account in the shaping of future government policies. But the underlying nature of those policies was not changed.

The period beginning in 1968 thus saw a short-term swing back toward increased state activity to improve housing, at the same time as the longer-term underlying trend toward private sector primacy continued to make itself felt. Legislation protecting tenants from eviction (through the "Social Clause") was adopted in 1968. Although most rent controls were ended in 1970, vocal protest resulted in a new Tenant Protection Act *(II Wohnraumkündigungsschützgesetz)* in 1971. Incentives for worker households to save for home ownership were increased, in part as an anti-recessionary measure. The increase in rents in social housing, begun with the 1968 Amendments to the Housing Act, remained in effect, but increased housing allowances were made available to help pay rents in subsidized housing as well as newly freed private market rents by the Second Housing Allowance Act of 1971 (II *Wohngeldgesetz*). Even the level of housing construction under the First Subsidy System, of most benefit to lower-income families, and on a secular decline since 1955, took an upturn.

The key legislation of this period of conflict, however, was the Urban Development Act of 1971 *(Städtebauförderungsgesetz)*. The result of protracted negotiations, it expanded governmental powers to engage in those activities we call redevelopment and renewal in the United States. It provided funds for rehabilitation and neighborhood services, some protection for those adversely affected by renewal activities, and mandated decentralization of decision-making and citizen participation. The Act, on the one hand, resulted from long term developments in the economic sector (the development of tertiary activities, improved communications, the further concentration of economic activities) which made rationalization of inner-city land uses crucial to the economy as a whole and particularly to the business community. But, on the other hand, it clearly recognized the protests of those displaced by earlier local renewal efforts or faced with escalating housing costs as a result of such activities.

PHASE III — FINE-TUNING COMPROMISE

In the latter part of the 1970s, with the apparent quiescence of the conflicts of the late 1960s, the attempt to streamline, rationalize, codify, develop — generally to fine tune — the basic instrumentalities of earlier housing policies was the main focus of governmental attention. Formal citizen participation mechanisms, in part modeled on experience in the United States, helped to direct resident reactions into nonconfrontational channels. The overall level of new social housing units completed in the First Subsidy System began to decline again as of 1973, and went from 169,336 units that year down to 40,239 in 1979. The Tenant Protection Act of 1974 weakened the provisions adopted in 1971, but kept some protections for tenants. The revision of the Federal Building Act in 1976 streamlined governmental action in the field of planning and land-use controls. In 1974, major funding was provided for the first time in support of private rehabilitation. Given significant impetus by the pressures of fiscal and monetary policy as a stimulus to employment and economic recovery, the shift in the emphasis of governmental support from new construction to rehabilitation has remained in effect. The 1977 Housing Rehabilitation Act incorporated these programs into a single statute, and streamlined some of the administrative policies begun in 1974. A series of studies and experiments since then have tried to overcome some of the shortcomings revealed in the original program, and inner-city renewal and housing rehabilitation remain a concern of current policy, with ongoing incremental changes — fine tuning — still required.[11]

Present trends in West Germany are converging very closely to many continuously prevalent in the United States, discussed below. The private market has the dominant role, and the gradual withdrawal of governmental support from the housing sector will continue — even if its ultimate levels, for political reasons, are unlikely to reach the uniquely low proportions being reached today in the United States. The employment-generated need for new housing for lower-income households has passed; the political pressures for it have diminished. The specter of tax resistance will make it easier to withstand what political pressures do exist. A general retrenchment in the level of governmental social expenditures is already evident. Changes arising out of the geographic redistribution of economic activities (although on a much lesser scale than in the United States), changes in the spatial requirements of industry and the secondary and tertiary sectors, and changes demanded by some of the costs produced by uncontrolled suburbanization and sprawl, e.g., energy costs, will all lead to an increasing focus on inner-city problems.

Up to now, there has been an apparent harmony of interests in dealing with these problems, once the initial wave of clearance and protest subsided. However, there is reason to believe that this harmony will not continue for long. The differential interests of low- and

high-income groups as to housing, of employers for sites as against
real estate interests for maximum appreciation, and conflicting
pressures for governmental expenditures and for tax-reducing
economies, will all make themselves felt. The types of conflicts
around issues of displacement, rising costs, low vacancies, and restricted opportunities, which are visible in many parts of the United
States today, are already being felt in the German Federal Republic.
The housing of foreign workers has never been satisfactory; it is now
contributing to making older inner-city areas the source of growing
concern.

Evidence of disharmony is beginning to appear more widely. Its
most visible manifestation is the squatter movement: the occupation
of empty houses illegally, sometimes spontaneously, defiantly, and,
most importantly, under a claim of right and of justice. Not only West
Berlin and Cologne and Freiburg, but over seventy cities throughout
the country have seen some form of squatting take place in the last
two years. What is remarkable is not only the extent of the movement,
but the widespread support and sympathy it has aroused. The buildings being squatted are in all cases empty ones, being held off the
market for speculation, for modernization, for demolition, or for
conversion to other uses. That such a situation should exist at a time
of widespread housing shortage appears absurd to a broad spectrum
of people. So the squatting movement represents a resurgence of
opposition to prevailing policies not only through the actual squatting
(which is in real numbers quite limited), but also because its acceptance and support indicates widespread recognition of the shortage
and the need for action. In some places, there has been some accommodation to it (West Berlin, Cologne); in others, it has been met with
direct and forceful police action (Freiburg). For the first time since the
early 1970s, there are direct and even violent confrontations on housing issues.

Clearly, other things are also fueling these confrontations: the
Alternative Movements, the ecological demonstrations, the antinuclear movement, the youth revolt. But housing seems to be playing a
central, if symbolic, role in these new events. The compromises
worked out in the late 1960s and early 1970s among the private housing
industry, business interests as a whole, and the users of housing are
now showing strains at the seams. The Social Democratic Party, with
its labor constituency, has not yet decided how to respond to the
strains. In Berlin, where the squatting movement is probably the
strongest, it responded with concessions when it was in power; in
other places (and in West Berlin too, many feel) its underlying sym-

pathies were not at all with the squatters. Given the general direction of government and the economy, fiscal pressures, the threat of inflation, and international pressures, the users of housing may have increasing cause for protest in the near future — as indeed is also the case in the United States.

HOUSING POLICIES IN THE UNITED STATES

The history of United States housing policies is fundamentally similar to that of West Germany, but differs from it in two important regards, each of which has a specific manifestation in the postwar period.

First, German economic growth has been dependent on an intimate involvement of the state with the details of economic coordination from the time of Bismarck on. The U.S. economy, by contrast, industrialized earlier with less need for governmental direction. With a continent on which to draw for resources and markets, and a size and power permitting the almost uninterrupted extension of international domination, the United States has not required as extensive state involvement. Both the fact and the ideology of unrestrained privatism[12] thus have been far stronger in the United States than in West Germany, the reality and the acceptability of a larger explicit state role, far less.

Second, German economic and political developments have produced a strong, self-conscious, and often militant working class and working-class political organizations. The United States, by contrast, has had much more wealth more easily obtained, vast expansion possibilities (at home and abroad) which made individual mobility an attractive goal for individual workers, and an internal history free of the legacy of a struggle against feudalism. No powerful working-class political organization or ideology moving radically to affect the benefits of governmental action has therefore arisen in the United States.

Obviously both these theses need vastly more extended discussion than they can be given here. Neither is new; the crucial point here is that *both* political *and* economic differences underlie the differences in housing policies, just as deeper similarity lies under both political and economic processes.

The immediate postwar context for these two differences are: First, on the economic side, the German economy lay shattered after World War II, and required a centrally planned concentration of resources for its renewed growth. The U.S. economy after the war, by contrast, was at a peak of productive power, facing a problem of over-capacity rather than its opposite. Second, on the political side, the West German leadership chosen by the Western allies after the war faced a serious problem of political legitimacy, as a consequence both of the Nazi past and the present alternative model of East Germany. In order to ensure itself the allegiance of its own citizenry, it adopted significant social programs. The leadership of the United States took a different political postwar tack. The potential attractions of a radically different alternative were submerged by the cold war ethos and McCarthyite repression rather than extensive social programs. When confidence in the stability of the system was shaken in both countries in the mid-1960s, their reaction was much more alike, because by then both economic and political conditions were much more similar.

DEVELOPMENT OF U.S. POLICY

The U.S. experience can be more briefly recited, because it is more familiar. The historical narrative divides roughly into six periods, which can be summarized as follows:

(1) Postwar shakedown (1945 to 1949)

(2) Suburbanization (about 1950 to about 1964)

(3) Conflict and Revolt (about 1964 to about 1968)

(4) Transition (about 1968 to 1972)

(5) Reaction (1972 to ?)

(6) Resistance and Renewed Conflict?

The initial postwar years represent a fascinating study in "policy making," for they are a period in which the explicit debate on housing policies was more vigorous, deeper, and more varied than at any other period in the nation's history (with the possible exception of the early New Deal or the period of conflict in the mid-1960s). Yet what actually happened was only marginally affected by the debate, and much more determined by the evolution of non-housing-oriented political strategy and macrolevel economic forces.

The debate was so vigorous because the war had unleashed a set of heightened expectations and had given political access to a set of participants whose roles and power were new and untried. The political stance of the returning veterans, the influence of the wartime alliance with socialist countries, the political ferment in Europe and throughout the world, the movement of blacks to urban areas and into the essential industrial labor force, the availability of resources at a level not approached since before the Great Depression to deal with issues not tackled for fifteen years — all of these created a tremendous and vocal ferment, including debates on housing policy both inside and outside the halls of Congress.

The debate concerned a number of issues that would appear central to housing policy: the nature and extent of public housing; the attitude toward redevelopment, slum clearance, and rehabilitation; the mechanisms of housing financing; policies toward racial segregation; and generally, the nature and extent of the state's role. The Housing Act of 1949 seemed to be the resolution of these debates. It reestablished and refunded public housing; it started (and created the legal mechanisms for) redevelopment of central city areas and slum clearance; it built a vast Veterans Administration (VA) and Federal Housing Administration (FHA) mortgage insurance program aimed primarily at single-family owner-occupied housing. So it seemed to set major policy directions for some time to come, as a result of extensive and democratic deliberations.

Yet the fact is that the major direction of policy, and the actual results of the 1949 Act, were set by quite different factors. The two major ones were, first, on the economic front, the need for a major expansion of the internal market to meet the productive capacity and to absorb the buying power built up during the war. The wave of suburbanization that took place in the 1950s was the result. No presidential economist, of course, canvassed the possibilities, chose this one, and then directed the state to underwrite and stimulate it.[13] Rather, the forces whose interests were met by this direction for policy were so strong, the resistance so weak, the momentum so self-reinforcing, that all fell before them. The private housing industry, from builders to financiers, from land developers to craft unions, were well served by the trend; only central city interests did not enjoy a piece of the action. Their efforts to get into the game dominated much of the conflict around housing policies for the next two decades, at the end interacting with even more powerful forces — economic changes in the nature of business activity and political changes in central city constituencies — to produce the conflicts of the mid-1960s.

The second major force actually determining the events of the post-1949 period was the political strategy chosen by the dominant leadership to deal with the aspirations and alternatives evoked by the war. Winston Churchill's Fulton, Missouri, speech sounding the keynotes for the Cold War as international policy, and Joseph McCarthy's unleashing of political repression at home, represented the decision to put down, rather than conciliate, potential opposition. The pie seemed big enough and growing rapidly enough to buy off the veterans and many blacks; the level of funding of public housing, more generally the resources committed to pursue the goal of "a decent home in a suitable living environment for every American family" so ringingly proclaimed in the 1949 Act, did not have to be very high if this was the approach. The carrots were VA mortgages, rising incomes to support suburbanization, and some improvement in absolute living conditions for minority groups; the sticks were political repression of the left and fear-sustained integration of the moderates. The combination worked, and large state-assembled resources did not have to be progressively distributed to achieve the goals. Eisenhower fell into the popularity Adenauer had to work to achieve.

Central city interests did not initially come out well in this process. The redevelopment provisions of the 1949 Act were limited in scope, clumsy in execution, sparsely funded. The tie-in to new housing provision was unwelcome; what downtown interests wanted was the means to clear land of unprofitable uses and assemble it for more profitable ones. They wanted to take advantage of pent-up demand, and to be able to adapt to changes in the nature of that demand because of shifts in modes of production and consequent spatial needs.

The 1954 Housing Act, with its much more flexible provisions for urban renewal and its much greater funding, provided the answer. In general, as Albert Cole,[14] one of the Act's chief architects, later described it, "The legislation was enacted with the purpose of placing greater reliance on private rather than governmental action" (quoted in Fish, 1979: 285). For central cities, it "was designed to place the major responsibility for development upon the private sector." It gave statutory support to the idea that urban renewal should contribute to improved housing; one of its requirements was that each city requesting funds would specify intended housing improvements in a "workable program." But the "predominantly residential" requirement was one of a number of loopholes permitting the result to vary substantially from the legislative declaration: Areas being cleared were only required to be "predominantly" residential, and even this link to the goal of improving housing was interpreted flexibly to favor commercial redevelopment. Finally, a specific exception of 10 percent was provided from the residential requirement, and the figure was gradually increased till it reached 35 percent in 1965. It was expanded even further by administrative manipulation (astute drawing of boundaries, counting questionably residential units). Pressures from the industry steadily weakened the housing link of urban renewal until it became transparently a central business district development scheme. Only strong countervailing pressures from a different direction changed the situation.

CONFLICT AND REVOLT

The events that led up to the widespread unrest of the mid-1960s have been well recounted elsewhere.[15] Direct community resistance to the displacement caused by urban renewal was a significant part of the unrest. Intellectuals — Chester Hartman, Herbert Gans, Staughton Lynd — played a creditable role in bringing the facts to public attention, but it was the changing composition and political stance of central city residents that produced the policy responses of the mid-1960s. The civil rights movement and the ghetto revolts, from Watts in Los Angeles in 1965 to Detroit in 1968, shook up the establishment. Despite earlier warnings, it was massive resistance and unrest that produced the Civil Rights Act of 1964, the Demonstration Cities (later, for obvious reasons, renamed Model Cities) Act of 1966, and the comprehensive Housing Act of 1968, with its expanded funding for assisted housing construction and rehabilitation; development of new programs to affect groups and situations not earlier touched; changes in financing mechanisms, guarantees, and subsidies to reach lower-income households. The establishment of the cabinet-level Department of Housing and Urban Development (HUD) in 1965, to replace the lower-level Housing and Home Finance Agency, symbolized the change in leadership attention to housing problems.

But the unrest of the 1960s never took solid political form. It was, of course, political in the deepest sense. It dealt directly with governmental decision-making processes, and challenged the distribution of political power. But the challenge was negative; no alternative organized force was generated that tried to substitute itself for the holders of formal power. The ghetto revolts lacked organization or ideology of their own; their spontaneous character was both their strength and their weakness. The working class did not enter the fray directly, even though parts of it — public sector employees and their unions especially — benefited from it mightily. The more ideological left and youth movements that supported the resistance of the 1960s found themselves increasingly in the anti-Vietnam campaigns of the early 1970s, and because of weak organization never returned to their earlier concerns. So the unrest of the mid-1960s did not produce a powerful continuing force for change in the direction in which it seemed to be heading.

The first Nixon administration, in the transitional period from 1968 to 1972, revealed an uncertainty as to the real balance of forces on both sides. The inclinations of the dominant leadership were clearly toward a return to repression and a cessation of redistributive governmental programs in housing; yet the momentum generated by the earlier period was not that easy to stop, and the certainty of stability not that clear, at first. Thus, in the early Nixon years, despite an explicit conservative orientation, the housing activities of the state in fact were at their postwar peak (see Table 5.3). Economic developments — the recession of 1969, the competing pressures of the Vietnam war and the shortages they created, a weakened international economic position — perhaps contributed even more to the continuing federal presence in housing than did political uncertainties, for these specific years.[16]

REACTION

Nixon's safe reelection in 1972 signaled the solidification of a basic shift in strategy by the established political and economic leadership. The moratorium on all federal housing programs decreed by Nixon on January 1, 1973, was a statement that political unrest was no longer to be feared; if it appeared, it would be put down by whatever means necessary. Even by conventional U.S. standards, Nixon's political tactics were ruthless. To the extent that state housing programs constituted a buying off of a potential political opposition, they were no longer considered necessary. That interpretation still holds today, and stands in significant contrast to the attitude of the majority of the German leadership.

The 1974 Housing and Community Development Act represents a set of housing policies in which housing industry interests dominate more than at any time since 1954. Politically, the period is one of reaction; in terms of housing policies, the dominant note is privatization. Not that that note was new; state subsidies had always been a tiny portion of expenditures on housing, and the funneling of those subsidies had always been through the private sector. The greater "popularity" of public housing after the introduction of turnkey development in 1966, by which private developers rather than public housing authorities initiated and executed the production of public housing, is an example; the acceptance of Section 236 when its tax advantages for limited-dividend private developers became apparent

TABLE 5.3   Housing Construction in the United States, 1950-1978

(1) Year	(2) Total number of units started[a] (in thousands)	(3) Single-family units as percentage of (2)[a]	(4) Private housing with Federal finance as percentage of (2)[b]	(5) Low-income public housing as percentage of (2)[c]
1950	1,952	NA	34.7	0.01
1951	1,491	NA	27.6	0.7
1952	1,504	NA	28.0	3.9
1953	1,438	NA	28.4	4.1
1954	1,551	NA	37.6	2.9
1955	1,646	NA	40.7	1.3
1956	1,349	NA	34.1	0.9
1957	1,224	NA	24.3	0.9
1958	1,382	NA	28.8	1.1
1959	1,531	80.3	28.9	1.4
1960	1,274	77.5	26.4	1.3
1961	1,337	71.9	24.5	1.6
1962	1,469	66.2	22.9	1.9
1963	1,635	62.0	17.9	1.7
1964	1,561	62.3	16.9	1.6
1965	1,510	63.9	16.3	2.0
1966	1,196	65.2	16.4	2.6
1967	1,322	63.9	17.5	2.9
1968	1,545	58.3	24.8	4.7
1969	1,500	54.1	18.9	5.2
1970	1,469	55.5	30.5	5.0
1971	2,085	NA	29.8	4.4
1972	2,379	55.1	20.0	2.5
1973	2,057	55.1	12.0	2.6
1974	1,352	65.8	12.4	3.2
1975	1,171	76.5	14.9	2.1
1976	1,548	75.3	15.8	0.4
1977	1,990	73.0	15.5	0.3
1978	2,023	71.0	15.1	0.5

SOURCE: HUD Statistical Yearbook, U.S. Bureau of Census: Construction Reports, Series C20.

[a]After 1962 includes farm housing.

[b]FHA and VA, which are the main long-term federal programs.

[c]Column 5 refers to units made available for occupancy, whereas columns 2, 3 and 4 refer to units started.

is another. But the 1974 Act placed the private sector formally on the throne, even if it was not until the Reagan administration that it received its full purple robes, crown, and scepter.

The Community Development Block Grant (CDBG) program and the Section 8 housing program were the two major contributions of the 1974 Act. The former represented a limited political compromise with more urban-based leadership and constituencies. While the geographic areas received funds under a formula that reflected population, poverty, and age of housing stock (thus presumably reflecting need), the distribution *within* areas, study after study has shown, was much more regressive.[17] The survival and probable ascendance of Urban Development Action Grant (UDAG) type funding, by which only those projects are subsidized that can also attract private capital, is symptomatic of the same trend. Working-class, minority, youth, and elderly pressures have always been more effective at the federal level in the United States than at the state or local levels, in all but a few communities; the shift to decentralization represented by the CDBG program, and in general supported by Nixon and ongoing through Carter and Reagan, is thus itself a regressive shift.

The Section 8 housing program is the closest the United States has come to a housing allowance program. Its existing housing component, under which subsidy is extended for the rental of existing private units, is almost a pure demand-side subsidy. Even its arrangement for the financing of new construction and rehabilitation gives maximum scope to the private sector: ownership of the subsidized housing, management, tax benefits, choice of location, size, occupancy, are all, subject only to limited government supervision, left in private hands. While this explicit housing allowance program, tentatively supported as an experiment by the Nixon administration, could function with minimal public control over the private market, it lost support primarily because of the problem of limiting its coverage. Funding for supply-side programs can be easily limited; projects are considered administratively, and when funding runs out, no new projects are built. But a housing allowance program is logically an entitlement program, like welfare. All those eligible must receive benefits; how could one set a cutoff in numbers? One could hardly go by priority of application without having a riot on the lines waiting at the door to file. Even with Section 8 certificates this has occasionally been a problem.

It may well be that the 1974-1975 recession halted the next logical step in the direction of reprivatization — the absolute termination of progressive federal subsidies in housing. But, even under the

TABLE 5.4   Distribution of Governmental Housing Benefits, by Income, United States[a]

(1) Income Group	(2) Income Group as Percentage of all Households	(3) Income Group as Percentage of all Households Receiving Benefits
Under $5,000	22.5	9.8
$5-$10,000	20.3	6.0
$10-$20,000	28.7	24.6
$20-$50,000	25.8	52.1
Over $50,000	2.7	7.5
Total	100.0	100.0

SOURCE: Cushing N. Dolbeare, National Low-Income Housing Coalition, Washington, D.C., May 1980.

[a]This table differs from Table 5.2 in the formulation of column 3, which is by percentage of households, not percentage of benefits. The latter figures would probably show the distribution as even more regressive.

presumptively more liberal Carter administration, levels of funding were reduced. The Reagan administration is now simply going much farther and much faster down a road that had been entered at least eight years earlier. The political pressures for progressive housing policies have been written off as not worth attention. Much of the housing industry — property owners, flexible lending institutions, speculators, realtors — is benefiting from inflation and doing quite well without further federal assistance. Builders, hurt by high interest rates, look to that issue, and to changes in financing mechanisms, as the way to increase their profits; the cost of a federal program to build for low-income households, given both high interest rates and the political complexion of the administration, seems too large to be worth an effort.

THE CURRENT SITUATION

The present course, therefore, is reduction of the federal role absolutely, and the restriction of what remains to a supporting role for the private housing sector. The regressive distribution of those benefits that continue is shown in Table 5.4. National economic conditions determine the allocation of resources to the housing sector as a whole (largely through the mechanism of interest rates); the housing industry is subordinated to larger business interests. Within the housing industry, the state's role assists suppliers, with fewer and fewer benefits to the presumptive beneficiaries of a state whose claim to benevolence is more and more explicitly repudiated.

Resistance, of course, exists, and is slowly increasing. It has not, however, reached the levels that might have been expected, perhaps for three reasons. First, the United States displays the perennial weakness of that potential but inchoate opposition which in Germany is well-organized and traditionally respected in ruling circles. Regardless of how shallow the differences are in Germany between the Social Democratic Party and the trade unions on the one side and the Christian Democratic Party and the leadership of business on the other side, the SDP represents a powerful liberal force with a large constituency having a solid stake in the welfare state. To the left of it (and partly within it) is a diverse, articulate, and unpredictable set of movements with an explosive potential to undermine at least the current leadership of the state. Neither exists at a similar scale in the United States.

The second reason for the as yet limited protest may be the sharp edge which the administration has thus far succeeded in inserting between the "middle class," including a large segment of the working class, on the one hand, and the poor, the unemployed, minority group members, many women and elderly and young people, on the other. Where the interests of moderate and higher-income people have been affected by the same problems that affect poor people, then poor people's problems have also been addressed.[18] That situation has, for the time being, not risen to the level of widespread awareness among the taxpayers, urban residents, and working people of the United States.

The third reason for the weakness of protest at what appear to be massive cuts in state benefits in the United States is the least certain, but perhaps the most important. State housing benefits for poor people may in fact never have been that significant anyway. At no time in U.S. history did more than 5 percent of the country's population benefit from the progressive aspects of governmental housing programs. Even that 5 percent received niggardly benefits, at the cost of social stigma, segregation, bureaucratic dehumanization, loss of control, and surrender of self-respect. It is no utopia that is being shattered; there is simply a little less meat in an already thin soup kitchen. Anyone with any choice is already eating elsewhere. The state has never functioned, in the housing field, for the benefit of the needy. Many of the needy now hardly notice its withdrawal.

STRUCTURAL EXPLANATIONS

The chronological account of housing policies in the United States given above deals with middle-range time spans. Underlying, invariant trends, continuous and crucial in their implications, are not picked up in looking at changes from year to year or even decade to decade. They may be listed as follows, returning to the schema outlined at the opening of this chapter:

(1) Housing sector investment has been determined — more than by any other single factor, more than by user needs, more than by supplier pressures or grass roots demands, more than by shifts in political party or intellectual trends — by trends in the private capital market. Interest rates and returns on equity regulate investment and expenditure activities in a market economy so as to maximize overall profit levels. Changes in interest rates are particularly effective in the housing field because of its heavy dependence on borrowing.

The pattern in the United States, as opposed to Germany, has been to avoid state interference with fluctuations in the housing sector caused by fluctuations in economic activity, and in interest rates in particular. *Kunjunkturpolitik,* the management by the state of economic cycles, very much an objective of German economic policy, is underdeveloped in the United States, for reasons sketchily outlined earlier. The difference in the two countries, however, lies in the *extent* to which the state is used for the purpose of guaranteeing profit maximization, not in *whether* it is so used.

Within this broad economic determinant, there is significant variation, and we have seen its effects in the brief history set forth above. Immediately after the war, the support of consumption placed a significant call on U.S. state efforts; by the 1960s, that factor (and the level of federal mortgage insurance efforts, for instance) played a much smaller role. Today, restraints on consumption appear the immediate goal of much federal policy. The control of inflation, the stimulation of capital investment elsewhere, international economic relations, all modulate the underlying theme of support for economic growth; the primacy of that support remains constant.

(2) Similarly, in the political sphere the interests of stability are ongoing and permanent, although the response to particular challenges may be short-term and variable. In the permanent category fall policies at the two ends of the housing spectrum: the emphasis on home ownership, and the stigma attached to public housing. Together with state policies influencing racial segregation and discrimination,[19] homeownership and public housing stigma operate to stratify, harden, internalize social differentiation. For those whose loyalty to the system may be in question, such policies create a stake, a reward, if they behave; a punishment if they do not. Housing subsidies may be expanded or contracted as the perceived need to coopt opposition rises or falls; the support for home ownership, the stigma attached to

public assistance, and policies sustaining racial differentiation, remain underlying motifs throughout.

(3) The housing industry, and certain user groups (ethnically, geographically, or economically defined) are the most visible influence on housing policies much of the time. As the New York *Times* once commented:

> Traditionally, housing bills have been an uneasy compromise between the demands of the private real estate industry and the moral claims of the slum-dwellers, the elderly and the minority groups, with Congress usually more responsive to the former than to the latter.[20]

Its comment happened to be on the 1964 Act; by 1968 the "moral claims" had been reinforced by physical actions, and the result was different. But the underlying truth of the observation remains.

The German picture differs from that of the United States in that special interest groups play a much smaller role; the difference, however, is superficial. At the formal level, the strength of party organization in Germany makes pressure at the level of the legislative process of secondary usefulness; relations with the party leadership play a larger role than in the United States. More fundamentally, because the role of the state itself is larger in Germany, it can less be permitted the kind of uncertainties and shifts from year to year that characterize many aspects of U.S. housing policies (Gough, 1975). Too much is at stake; a more disciplined process is required. But the process takes into account the same set of interests as in the United States. The difference lies in the form and level and smoothness of integration between the state apparatus and private interests, not in the critical linkage between them.

CONCLUSION

The differences between German and U.S. housing policies in general, then, are more in form and quantity than in substance or direction. In both, the underlying commitment is to the private market. But in both, the private housing market requires constant state involvement; in neither can it function autonomously. In both, the state's day-to-day role is determined primarily by the needs of the private housing industry, tempered by other varyingly influential business and user groups. In both, the longer-term nature of the state's role is determined by national requirements for profit maximization in the economy and for stability in the political system, which take priority over narrower sectoral housing interests when there is a conflict. Postwar history in both countries has shown how these basic interests — housing industry prosperity, general business profitability, and political stability — have shaped current policies, with the housing industry playing a larger and larger role as the more general economic and political demands on the housing sector are met. In neither country have the needs of the users of housing been the major determinant of the state's policies. And in both, the advances of the last 35 years seem to have run their course, and the prospects for the future include increased privatization, increased hardships, and increased conflict.

NOTES

1. "Germany" is used hereafter in lieu of "West Germany," or the more formally correct "Federal Republic of Germany," for the sake of readability. It is always intended to mean the Federal Republic if the reference is to the post-World War II period, and Germany proper if in reference to pre-1945 events.

2. See Bryant and White, 1976; Burns and Grebler, 1977; Duclaud-Williams, 1978; Dumouchel, 1978; Fainstein and Fainstein, 1978; Fuerst, 1974; Headey, 1978; Mandelker, 1973; Pugh, 1980; Wendt, 1962; and the writings of Michael Harloe, Jay Howenstine, and Suzanne Magri.

3. See my review essay of several current works in the field in *International Journal of Urban and Regional Research*, Spring, 1982. Comments on this chapter would be very welcome indeed; they may be addressed to the author at Avery Hall, Columbia University, N.Y. 10027.

4. A more detailed, but very tentative, formulation of what follows may be found in P. Marcuse, "The Determinants of Housing Policy," Working Paper 22, Columbia University Division of Urban Planning, 1981, 50 pp.

5. "Policies," not "policy." The phrase "housing policy" already implicitly accepts the notion that there is a single housing policy, a single drive or direction or objective which the state autonomously pursues — precisely the point the present paper finds false. See also Marcuse, 1978.

6. Of course much more than housing itself is involved: transportation, recreation, sanitation, safety, public facilities, and services. There has been a fruitful, if sometimes excessively jargon-laden, discussion around the concept of collective consumption since the term was first put into use by Manuel Castells, 1977.

7. The extended or social reproduction of the labor force, as well as its physical reproduction, is here intended. The phrase "reproduction of the labor force" is potentially confusing in dealing with these two linked concepts, and is in addition jarring in its special use of words with quite different connotations in ordinary English usage, so it is avoided here.

8. Although not inevitably. Most of what government can do can also be done privately, given the incentives: zoning in Houston; land-use planning in Columbia,

Maryland; public transportation in the airways; emergency housing for war workers have been done as successfully privately under certain circumstances as publicly under others.

9. Apart from interviews and official sources, the following account relies on Pergande, 1973, and Institut Wohnen und Umwelt, 1980; and on discussions with, or reading the work of Ruth Becker, Adalbert Evers, Hartmut Frank, Tilman Harlander, Margit Mayer, Eberhard Mühlich, Ilona Mühlich-Klinger, Konrad Stahl, Rudi Ulbrich, and Helmut Wollmann. Many of them, however, might disagree strongly with some or all of the interpretations suggested in what follows.

10. By Rudi Ulbrich, now of the Institut Wohnen and Umwelt, Darmstadt.

11. At present writing, (fall 1981), the federal modernization program seems scheduled for dismantling through budgetary curtailment.

12. There are of course many sources that could be cited for this contention, but Warner's *The Private City* (1968) and the work of Douglas Dowd, David Gordon, and James Weinstein all provide solid detail.

13. Although Truman was quite explicit about the role of housing: In his message to Congress of September 1945 he said, "The largest single opportunity for the rapid post-war expansion of private investment and employment lies in the field of housing, both urban and rural. . . . Such a program would provide capital to invest from six to seven billion dollars annually. Private enterprise in this field could provide employment for several million workers each year. . . . A housing program . . . would in turn stimulate a vast amount of business and employment in industries which make house furnishings and equipment of every kind, and in the industries which supply the materials for them" (Barton and Matusow, 1966: 92).

14. Cole was, at the time of passage of the 1954 Act, administrator of the Housing and Home Finance Agency, the predecessor of the Department of Housing and Urban Development.

15. Among the best accounts is that of Cloward and Piven (1972).

16. I am indebted to Emily Achtenberg for stressing this point, and warning of the dangers of looking at only one determinant of policy (be it political or be it economic) as decisive at any point. And the two are of course directly related to each other. It is interesting to speculate that the difference in reaction to economic problems in 1970-1972 compared to 1975 or 1980 reflects to some extent the difference in the nature of the economic problems, but also reflects a shift in the dominant strategy believed best able to cope with them, i.e., a shift in analysis.

17. See the work of the Center for Community Change's Community Development Block Grant Monitoring Project (including Marcuse, Medoff, and Pereira), *Triage: Programming the Death of Communities,* and of the Brookings Institution, and even HUD's own annual reports on the program.

18. See Piven and Cloward, 1977; and Marcuse, 1981.

19. See National Committee against Discrimination in Housing (1967), and subsequent bulletins and publications of the Committee.

20. *New York Times,* September 18, 1964, as quoted in Gelfand (1975: 361).

REFERENCES

BARTON, J. and A. MATUSOW [eds.] (1966) The Truman Administration: A Documentary History. New York: Harper & Row.

BRYANT, C. and L. WHITE (1976) "Housing policies and comparative urban politics: Britain and the US," in J. Walton and L. Masotti (eds.) The City in Comparative Perspective. New York: John Wiley.

BURNS, L. and L. GREBLER (1977) The Housing of Nations. New York: Macmillan.

CASTELLS, M. (1977) The Urban Question. Cambridge, MA: MIT Press.

CLOWARD, R. and F. PIVEN (1972) The Politics of Turmoil. New York: Vintage.

DUCLAUD-WILLIAMS, R. (1978) The Politics of Housing in Britain and France. London: Heinemann.

DUMOUCHEL, J. (1978) European Housing Rehabilitation Experience. Washington, DC: National Association of Housing and Redevelopment Officials.

FAINSTEIN, S. and N. FAINSTEIN (1978) "National policy and urban development." Social Problems, 26 (December).

FUERST, J. [ed.] (1974) Public Housing in Europe and America. New York: John Wiley.

GELFAND, M. (1975) A Nation of Cities. New York: Oxford University Press.

GOUGH, I. (1975) "State expenditures in advanced capitalism." New Left Review 92.

HEADEY, B. (1978) Housing Policy in the Developed Economy. London: Croom Helm.

Institut Wohnen und Umwelt (1980) Grunddaten, Darmstadt.

JACOBS, M. (1976) "The political economy of public housing." Master's thesis, Columbia University.

MANDELKER, D. (1973) Housing Subsidies in the United States and England. Indianapolis: Bobbs-Merrill.

MARCUSE, P. (1981) "The political economy of rent control," in J. Gilderbloom (ed.) Rent Control: A Source Book. San Francisco: Foundation for National Progress.

————— (1978) "Housing policy and the myth of the benevolent state." Social Policy (January-February).

National Committee against Discrimination in Housing (1967) "How the federal government builds ghettos." New York.

PERGANDE (1973) Die Gesetzgebung auf dem Gebiete des Wohnungswesens und des Städtebaues. Deutsche Bau — und Bodenbank.

PIVEN, F. and R. CLOWARD (1977) Poor People's Movements. New York: Random House.

PUGH, C. (1980) Housing in Capitalist Societies. Farborough, Hampshire, U.K.: Gover.

U.S. Department of Commerce (1980) Statistical Abstract of the United States. Washington, DC: Government Printing Office.

WALTON, J. and L. H. MASOTTI [eds.] (1976) The City in Comparative Perspective. New York: John Wiley.

WARNER, S. (1968) The Private City. Philadelphia: University of Pennsylvania Press.

WENDT, P. (1962) Housing Policy — The Search for Solutions. Berkeley: University of California Press.

Part III

Regional and Urban Development:

Less Developed Countries

6

The International Economy and Peripheral Urbanization

JOHN WALTON

☐ THIRTY YEARS AGO Oscar Lewis published his famous case study of "urbanization without breakdown" beginning a revolution in research on Third World cities (Lewis, 1952). Prior to this work focused on Mexico City, urban settings in the underdeveloped countries had received little attention, save for misgivings about the sundry forms of social disorganization that were assumed to follow from moving too fast and too far along the rural-urban continuum. Following Lewis's essay urban anthropology became a respectable pursuit and a growing fund of information appeared on the intricate ways in which rapidly urbanizing societies become very organized indeed. In many ways Lewis created a paradigm whose implications are still being developed.

Once anthropological work in cities became respectable and the study of humankind extended to society and economy, theoretical perspectives began to supplant descriptive case studies and to mature apace with empirical discovery. Separate accounts of this theoretical evolution have much in common. McGee (1979: 49) suggests that the initial perspective "portrayed the urbanization process in the Third World cities as similar." Later, the western model was superseded by various dualistic models based on rural and urban sectors, traditional and modern societies, or formal and informal economies. Currently most analyses subscribe to the general tenets of dependency theory in which various sectors are interdependent and closely linked to global capitalism. Elsewhere I have posited four theoretical approaches to

urbanism that understand the city as a market system, a mechanism of
social integration, a center of social pluralism, and as cause and
consequence of social development (Walton, 1981a). Doubtless there
are other versions of the "stages" through which our thinking about
urbanization has passed, but I expect that all of them have in common
the idea that naive notions of some historical continuum gave way,
first to theories of modernization and later to now-fashionable ideas
about "urbanization under peripheral capitalism." I take that to be a
progressive trend, but I want to explore in this chapter some of the
cul-de-sacs of our seeming theoretical sophistication, and with that I
want to suggest some fresh ideas about direction.

THE THEORY OF PERIPHERAL URBANIZATION

The theory of urbanization under peripheral capitalism comes to
us from the intersection of several streams of social research. First is
the renaissance of urban studies in the advanced Western countries —
the new urban sociology (Walton, 1981b) that owes so much to the
work of Castells (1978), Gordon (1978), Harvey (1973), Lojkine (1976),
Pickvance (1976), and many others. Second is the parallel work on
Third World dependency (Frank, 1967; Cardoso and Faletto, 1969;
Oxaal, et al., 1975) and the related nature of dependent urbanism
(e.g., McGee, 1971; Roberts, 1978; Gugler and Flanagan, 1978; Portes
and Walton, 1976). Third is the overarching perspective on uneven
development in the modern world system that connects and promises
to extend the other streams (e.g., Amin, 1974; Wallerstein, 1974,
1980). Although these are the new theoretical influences which increasingly intertwine, judicious applications of peripheral urbanism
are also capable of incorporating earlier research from more conventional perspectives — of building on solid work here interpreted in
a more holistic theoretical framework (Harvey, 1973; Walton, 1981a).

Despite its recent beginning, the study of urbanization under
peripheral capitalism includes an impressive amount of research and
some well-established tenets. I will summarize these briefly in an
effort to suggest the work's flavor, scope, and representative contributions, without pretending to offer a complete codification. The
study of peripheral urbanization starts, of course, with the axiom that
Third World development is constrained by and interacts with the
global political economy. Moreover, the very process of urbanization
is a critical dimension (schematically, a cause and effect) of uneven
international development. That is, in some ways the rapid urbanization of the Third World is *caused* by the expansion and penetration by

the global economy in the form of commercialized agriculture for export, the destruction of traditional rural social organization, urban (push) migration, the exploitation of new urban markets of labor and import consumption, and the reorientation of the urban economy toward export production and trade. Similarly, urban underdevelopment (the problems of housing, services, unemployment, crimes of desperation, and so forth) are *consequences* of a local economy that is misdeveloped and deformed by its orientation to external demands and the profits of comprador classes closely linked to those demands.

The concrete workings of the peripheral urban political economy require careful specification. Many particular processes are closely related in a pattern of multiple and reciprocal causation that belies the singular language of cause and consequence. Although they can be isolated for analytic purposes, their actual movement must be understood in a broad dialectic. A logical first tenet of the theory of peripheral urbanization is that among the several important stimulants to primary urbanization (e.g., population growth, the centralization of political empires, interregional trade), participation in the global export economy is a systematically decisive correlate of increasing city size (e.g., Berry, 1964; McGreevey, 1971). Further, if unequal exchange within the international economy leads to sheer increases in the rate of urban primacy, it also skews previous patterns of urban hierarchy, or alters the "system of cities," and generates increasing centralization of activities within cities (Slater, 1975, 1979). When combined with the commercialization of agriculture and rural out-migration, themselves tied to the export economy, these changes disrupt regional patterns of self-sufficiency and interdependence. Similarly, the urban centralization of activities encourages greater ecological segregation within the city (e.g., Castells, 1972; Slater, 1978).

The first effect of these changes is an expanding urban population in a few cities such as the capital, provincial centers, or port cities that would be difficult to absorb in the occupational structure under "normal" (i.e., of indigenous growth) conditions. The rate of urban unemployment is likely to rise faster than that of urbanization. But, this general expectation is made worse by the special features of the new urban economy. The key problem is that industrial employment does not expand apace with population and export trade for several reasons. Unequal exchange is based on the export of raw materials with fluctuating demand and the import of costly manufactures with steadily increasing demand, especially among comprador classes which profit in this exchange. Foreign manufacturing firms set up shop in the national market through wholly owned subsidiaries, joint ventures, or companies capitalized completely from local sources though controlled abroad. In the more dynamic industrial activities these firms compete with national firms or prevent them from ever entering the realm of modern production. In more traditional industries (food products, textiles, and so on) foreign firms compete, buy up, and drive out local firms. The greater efficiency of foreign firms is based on economies of scale, tax advantages, and capital-intensive methods of production. All these imply a twin disaster for the local labor market — labor-intensive national competitor firms are eliminated and with them many jobs, while the new capital-intensive, foreign firms employ many fewer workers in proportion to the volume of production (those well-paid few they do employ constituting a "labor aristocracy" that also consumes more imports).

The massive and related effect of these developments is to generate an overcrowded tertiary sector of commerce, services, and public employment. Unemployment is a less appropriate description of this deformed labor market than underemployment and the steady expansion of the informal economy (Portes and Walton, 1981). These are some of the prominent structural conditions that produce the army of the underemployed who scratch out a meager subsistence in street vending, petty and personal services, artisan or low-paid traditional manufacture, and sundry clandestine (e.g., household production avoiding tax and labor codes), or illegal economic activities. All of these represent a *growing*, rather than vestigal, form of urban employment because they simultaneously provide the barest necessities *and* subsidize middle-class consumers and households as well as the casual labor requirements of the formal sector (Bromley and Gerry, 1979). A marvelous study by Birkbeck (1979) illustrates these complex linkages in the case of rag pickers at the garbage dump in Cali, Colombia, whose labors contribute to one of the city's largest industrial firms. Finally, the state endeavors to soften the contradictions of dependency and underemployment by expanding public jobs, and it often does this successfully in the sense of creating an unproductive, graft-ridden white collar proletariat, but at the expense of alternative forms of public investment (e.g., in urban services, housing, productive state enterprises).

The physical form and spatial organization of the peripheral city reflect and reciprocally recreate the urban economy. Obviously, the city is "overcrowded" with people earning minimal or no incomes. Slums and shanty towns proliferate without basic services. The state, endeavoring to meet the demand for low-cost housing and belated services literally essential for survival (e.g., potable water, sewage, medical clinics), nevertheless finds itself falling further behind the demand, given its meager revenues and obligations to public employment. Without overly complicating the picture, part of the state's fiscal dilemma is related to tax concessions to modern industry, the reduced number of taxable local enterprises, and the expensive infrastructure required by new industries, upper-class residential zones, and general services (e.g., transportation). As a result of all this the city becomes more segregated ecologically. The central city is increasingly devoted to commerce and administrative functions, expelling residences and mixed use, small-scale production activities. Upper-class housing and new industrial enclaves are located in choice surroundings convenient to transportation and services. Lower-class housing and small-scale production and commerce fill the interstitial zones and, frequently, a growing peripheral ring around the city.

This pattern of ecological segregation and unserviced slums is more than a residual consequence of a dynamic economy with other spatial priorities — more than a product of neglect. On the contrary, the slums and poorly serviced working-class zones are closely tied to the circulation of capital and profit. In these rapidly growing cities speculation in land and housing is a major economic activity, particularly attractive to local upper classes whose high profits from commercial activity lack other profitable avenues of reinvestment. The more profitable and expensive industrial firms, recall, are typically closed investments of the multinationals and their junior partners among local industrialists who provide the facilities for joint ventures. Why struggle to create new, productive industrial firms in competition with the multinationals when trouble-free investments in land and rentals will yield comparable profits? The rational response of local wealth is to engage in speculation which creates few permanent jobs (though many temporary ones in construction) and a growing scarcity of affordable land or housing. Speculation feeds inflation, driving the working class out of the housing market and into the growing slums. This is the essential connection between the modern sector of the economy and the massive problem of urban poverty, slum housing, ill health, and the host of vices forced upon people down to their last alternative.

Now, let it be clear that what I have provided in the last few pages is an overly schematic presentation of the theory of urbanization under peripheral capitalism. Similar expositions treat the more specific case of Latin America (Slater, 1978) and some of the measurable associations (e.g., export dependency, urbanization, industrial and service sector growth) assumed here to form a complex causal web (Kentor, 1981). I believe that the theory is a good one, that it is *useful* in sensitizing us to both the complexity of the whole and to critical links that may have more practical importance and, indeed, may vary widely in concrete circumstances. However, I also believe that this theory, like most, runs the risks of empty generality, hypostatization, mechanical quantitative "verification," and other blind alleys of research or political cynicism. The theory, in other words, should summarize where we stand now as a result of some valuable research and thereby serve as a starting point for fresh questions and practical policies.

THEORY AND PRACTICE

Unfortunately, the theory of peripheral urbanization and related developments in dependency theory seem to have done more to promote tediously repetitious research and scholarly ennui than innovative responses to the very circumstances that researchers have endeavored to document. In an odd manner by proving ourselves prescient we have made ourselves impotent. One typical irony is that research will uncover the formidable "structures of domination" that integrate and animate a problem like peripheral urbanism, concluding the otherwise forceful analysis with a set of "policy recommendations" which no sane party to the process (collaborator or revolutionary opponent) could possibly accept or implement. A recent, perhaps only momentarily jaundiced, statement by McGee (1979: 64-65) reflects the general tenor.

> For the moment, it seems that the greatest burden the Third World urban poor have to carry is the governmental elites who purport to be representing their interests, but instead are really collaborating with the overall processes in 'peripheral capitalist societies' which are encouraging the persistence of poverty and preventing any radical redistribution of income.

The statement appears under the heading of "alternative policies" and, like so many others, ends with a hollow urging of "radical" change. The root of this difficulty, I believe, is that the theory has been reified — it has surreptitiously become the reality rather than a device for organizing part of the reality. Once "heuristic assumptions" have become obdurate facts.

EFFECTS OF THE GLOBAL ECONOMY

Understanding this, as well as the original useful purpose of the theory, let us return to some first assumptions. For present purposes I will focus on just two of these. The theory of urbanization under peripheral capitalism assumes, correctly, that the global (or core) economy determines and constrains the course of Third World development. I would amend that to say that the global economy is one among several powerful influences on the development of third world cities — an influence that has direct, indirect, and remote effects, all of which interact with concrete local influences including social organization and cultural tradition. Second, the theory assumes that patterns of urbanism, both spatial and economic, reflect the interests or the requirements of the dominant "structure of capital." That I would amend to say urbanism is less a "reflection" of something else than a continuous struggle among various classes, status groups, and authorities in which domination is less the rule than negotiation.

The amended first assumption can be illustrated with a set of more and less direct effects of core capital and policies on the organization of the Third World city. Brief examples will be provided which could be vastly extended. Let us start with the most powerful *direct* effects. Suppose that a large multinational firm locates itself on the fringe of a moderately sized city and begins producing items for the local market, items which were available previously in a less copious or fancy form (think of slick packaged food products, metal or construction materials, chemicals of synthetic fibers). In the first instance the firm has probably been lured here because of the promise of the jobs it will provide, the local managers it will train, the capital it may bring, export to nearby regions or third countries, and even offer publicly on the local stock exchange at some time. The lure, of course, typically involves a tax holiday and the provision of the necessary infrastructure of energy, transportation, and so forth, all at public expense. On some of the promises the multinational firm does, indeed, deliver. Jobs are provided and they tend, in skilled categories, to pay better than the city average. But, in proportion to the volume of (capital intensive) production, the number of jobs provided is small and drawn from the ranks of the most skilled. The growing urban masses and immigrants are not affected at this stage. The equipment for the plant is probably imported, providing no stimulus or "linkage effects" on local industry. On the contrary, in most instances the firm's product competes with and often eliminates local smaller firms, creating a net increase in unemployment and a net decrease in profit realized at the local level (since much of the multinational's earnings are repatriated in open and covert ways). The firm's products sold locally are probably no better at satisfying a necessity (such as food or a building block) — they may be worse or they may be styled and priced as a luxury item. Commercially, the product is typically handled by sales monopolies or chain stores to the disadvantage of local merchants. Whether the product is for export or mainly local consumption, its distribution becomes the province of a comprador bourgeoisie.

As for the *indirect* effects, on one hand the multinational labor force that is skilled forms a new aristocracy of labor, while the larger (net) number of unemployed workers is pushed out of industrial jobs and into the tertiary sector or the informal economy. These "sectors" grow in numbers of people at the same time they decline in their share of the social product, which is increasingly hogged by monopoly industry and commerce. They "flourish" at the margin of subsistence by providing one another with petty trade and services and by subsidizing middle-class households and even large firms (as in the case of Cali's rag pickers) with cheap labor. But the disadvantages are spread beyond the working class. Middle-class consumers find their choices reduced and are lured into the stores of monopoly merchants. The latter, as suggested, realize grand profits, but reduced opportunities to reinvest them in productive enterprise. Their winnings typically go into luxury consumption (including expensive imports) and speculative ventures, particularly land and rentals, and banks at home or abroad. The intricate pattern of circulation, of course, is inseparable from another of production (analysts of similar processes vary in emphasis, but I doubt the claim of critics that many are "circulationists"). These indirect effects extend to the advanced countries in the results of multinational (conglomerate) expansion on patterns of consumption, employment, and class structure (Portes and Walton, 1981).

Third, there are *derived* effects of this process in the sense of changes in local policy and ambience that affect everyone. Most apparent in this category are changes in the spatial form of the city. New industries require expanded transportation facilities — new roads to adjacent markets or ports, new and larger airports, interurban transportation for workers (freeways and subways that level old houses and narrow streets). New zones of working-class housing, of industrial use, and suburban enclaves for the new managers are all commonplace. As noted, again, when these new urban works are mainly the responsibility of the state, scarce public resources are deflected from basic works in public housing (made more expensive by land speculation), potable water, electricity, sewers and drainage, paving, lighting, transportation to these unserved zones, and many more. As before with the indirect effects, these inconveniences are partly shared by the middle classes and affluent. Elegant neighborhoods in the central city are destroyed or made unpleasant or become too valuable as commercial property. New suburbs proliferate in the boom of real estate speculation. Finally, perhaps the most important derived effect is a reorientation of state policy toward the promotion of more (similar kinds of) industrialization, the public infra-structure to support it, and the social necessities to accommodate it (e.g., housing). There is a sea change deflecting the ship of state away from some appreciation of local needs and toward the allure of modernization.

Last, there are *contextual* effects best represented by changing consumer tastes for luxury products or status symbols and changing ideologies about the ends and means of development. State policy emphasizing modernization spreads to the enterpreneurial con-sciousness and popular standards of mobility and the good life. Some-thing best termed a pernicious "culture of modernization" develops in much the same way Veblen described the trend toward "pecuniary emulation" in the United States at the turn of the century. This is more than a preference for transistor radios, discos, and designer jeans — it is a reorientation of popular culture.

LOCAL AND NATIONAL INFLUENCES

Now, what we must appreciate is that all of these effects, including the direct ones — so much the exclusive emphasis of the theory of peripheral urbanism — take place within societies that also have strong political ambitions of their own, persistent patterns of social organization, and deep cultural traditions. A "peripheral urban place," seen from the ground up, appears as a city very much in its own right. Without looking hard one can locate the outposts of core capital in an Esso installation, the Goodyear factory, the twin scourges of McDonald's and Kentucky Fried Chicken (recently joined by Denny's, Burger King, and a host more), the omnipresent Coca Cola, and perhaps an IBM plant. But in the larger cities where these are attracted, they are enveloped in a larger set of institutions — many linked to the global economy, as I have argued, and many independently exercising their own influence on the shape of ur-banism.

Paralleling the effects of the global economy, a new politics of economic nationalism constitutes a potent local and national influ-ence — the "new international economic order" best characterized by the petroleum exporting countries, but also represented among producers of essential raw materials and food. As the struggle among advanced countries for Third World markets becomes keener, and as peripheral societies begin to appreciate their own growing leverage, new arrangements for the entry of capital and the terms of trade are being made. Protection of national industry is more possible, and greater concessions can be won from those foreign corporations that are allowed to enter (e.g., real increases in employment, decentralized location outside the cities where some urban migration may be stemmed, genuinely complementary industry that stimulates local linkages). These, of course, do not undermine global capitalism — they may assist it and modify it. But they do change the nature of dependency, perhaps even the extent. In most of the Third World (vigorously in Latin America and increasingly in Africa), dependency is far more than a privileged academic theory. It is the well-known enemy of economic analysts and policy planners — thanks often to academics and to the flow of liberating ideas. Naturally, the openings available to economic nationalism, as well as some of its own motives, vary widely across Third World countries. But for analytic (rather than sweeping descriptive) purposes the fact is important in two ways. One, it may be a trend. Two, it is an alternative, whether pleasant or not, that prevailing theory does not reckon. In a rush of optimism one could say it is an opportunity.

It is unfortunate that our most compelling theories of underdevelopment allow little room for social organization and cultural tradition. Indeed, these are often denigrated by a hard-bitten materialism that views anything "nonstructural" (whatever that means) as ephemeral at best and, at worst, deviously manipulative. This is a bit odd since in some of the same theoretical quarters accolades have, rightly, been awarded to the social insights of Barrington Moore (1966) and the analysis of working-class culture by E. P. Thompson (1963). In any case, I think it is a serious oversight that contributes to our theoretical failure to see beyond an unrelieved dependency and to our ennui.

Urban societies are not, anymore than real individuals, mere passive recipients of external determinants of their behavior. In straight economic terms they respond with new initiatives. The construction industry is a prototype. It both absorbs large quantities of labor (including engineers and architects) and stimulates a variety of local industries less easily penetrated by foreign capital (e.g., cement, wood, construction materials, furnishings). But broader cultural traditions are also alive. The central plazas, market places, walled cities, and "native" quarters persist sometimes as tourist attractions (an "industry" after all) and often as functional parts of the urban form.

Moreover, people adapt many of the products of the new urbanization to their own uses. Public transportation serves the local entrepreneur, down to the itinerant musician using a bus as a rolling stage. Public buildings are a focal point of vending and petty brokerage, including the sidelines that minor functionaries carry on from their official desks or doorsteps. As a regulator of this traffic, the doorman of the average government office building is a person of influence. In some ways, working-class culture controls, thrives, and gets more use from official quarters than the intended beneficiaries.

Beyond cultural preservation and adaptation a more important development is the production of new cultural practices and forms of social organization that are a part of rapid urbanization. The most overwhelming instance is the new urban culture of the lower-class housing and squatter settlements. Often these are created through invasion and neighborhood organization (Cornelius, 1975). The new "development" is planned in advance of occupation. Later streets and lots are laid out, committees organized for social control, and delegations appointed to beseech authorities for land titles and basic services. Although the settlements may be later eradicated, or simply ignored, more typically authorities do respond to community organizing efforts, since they represent the best available (perhaps the only) solution for accommodating the urban poor. And a low-cost solution at that. Grassroots organization of urban neighborhoods creates a lifestyle of its own, cooperative practices, a small-scale economy, and an experience of political socialization, as groups representing the new communities enter into negotiations with local bureaucracy and political parties.

Dour social theorists can learn much from poets like Neruda (1976: 58, 63-64), who describes the lower-class neighborhoods of Valparaíso, Chile:

> Valparaíso is secretive, sinuous, winding. Poverty spills over its hills like a waterfall. Everyone knows how much the infinite number of people on the hills eat and how they dress (and also how much they do not eat and how they do not dress). The wash hanging out to dry decks each house with flags and the swarm of bare feet constantly multiplying betrays unquenchable love. . . . I have lived among these fragrant, wounded hills. They are abundant hills, where life touches one's heart with numberless shanties, with unfathomable snaking spirals and the twisted loops of a trumpet. Waiting for you at one of these turns are an orange-colored merry-go-round, a friar walking down, a barefoot girl with her face buried in a watermelon, an eddy of sailors and women, a store in a very rusty tin shack, a tiny circus with a tent just large enough for the animal tamer's moustaches, a ladder rising to the clouds, an elevator going up with a full load of onions, seven donkeys carrying

water up, a fire truck on the way back from a fire, a store window and in it a collection of bottles containing life and death. . . . But these hills have profound names. Traveling through these names is a voyage that never ends, because the voyage through Valparaíso ends neither on earth nor in the word. Merry Hill, Butterfly Hill, Polanco's Hill, Hospital, Little Table, Corner, Sea Lion, Hauling Tackle, Potters'. . . . I can't go to so many places. Valparaíso needs a new sea monster, an eight-legged one that will manage to cover all of it. I make the most of its immensity, its familiar immensity, but I can't take in all of its multicolored right flank, the green vegetation on its left, its cliffs or its abyss. I can only follow it through its bells, its undulations, and its names. Above all, through its names, because they are the tap roots and the rootlets, they are air and oil, they are history and opera: red blood runs in their syllables. *

Let me reiterate that these new methods of urban social organization and cultural practice do not "counterbalance" or wash away the effects of peripheral urbanization. I am not trying to put a happy face on slum life, unhealthy squalor, and persistent poverty. I am trying to suggest the diversity and complexity of the situation, and even to suggest that it may bring forth some creative responses — none of which is found in the prevailing theory. Of course, there are analyses that would have us believe that slum life and the precarious informal economy provide solutions to urban poverty apart from dependency (e.g., Laquian, 1971; Hart, 1973). But there are other more sophisticated and holistic treatments that show the vitality of "marginal" communities (Perlman, 1976; Leeds and Leeds, 1976) and the important fact that many urban migrants and workers judge their life in the city as a positive improvement over what might have been (Balán, et al., 1973). I find it odd, at least, that "radical" exponents of peripheral urbanization who clearly side with the plight of the poor are at the same time almost indifferent to how those people evaluate their own circumstance.

CLASS STRUGGLE

Let us turn to the second ammended assumption, the idea that spatial and economic patterns of Third World urbanism are more than reflections of the dominant structure of capital — that they are the changing products of struggles among classes, status groups, and authorities. The huge point that is ironically ignored in much Marxist-inspired contemporary theory is that the very fundamental changes transpiring in the political economy are at the same time structurally determining new class alignments. To the extent, moreover, that these structural changes have been shown to prompt political organization (such as in the new squatter settlements and urban communities), new classes "for themselves" have formed and pursued characteristic methods of struggle.

In broad strokes, the class structure of peripheral urban economies embraces four categories: (1) domestic and foreign capital owners, senior executives, and state managers; (2) salaried professionals and technicians in public and private employment; (3) clerical and manual wage labor in public enterprises and private industry and services; and (4) casual wage labor, disguised wage labor, and self-employment in petty production and trade (Portes and Walton, 1981: 103). The latter category includes most of the urban poor and is the fastest growing. The transformation of the urban economy described previously has its most damaging effects on both categories three and four. New capital intensive production eliminates or deskills wage labor. More workers are forced into casual labor and self-employment of the informal economy, just as the opportunities available to that "sector" are reduced. Ecologically these two classes tend to be interspersed in public and private developments for the working class and in "irregular" settlements of squatters and leaseholders.

The labor protest of these groups may be organized around conventional issues such as industrial pay and conditions, consumer prices, and merchant strikes. More common and growing, however, are protests surrounding the conditions of collective consumption — transportation, urban services, housing, land titles, health, and education. This new pattern has several explanations. Conventional industrial action is reduced because there are fewer industrial jobs and those that exist usually pay well, because of employer paternalism (e.g., worker housing), and because a significant minority experiences occupational mobility. The more common form of economic protest surrounds the problems of costly food and rent. Often this is carried out by shopkeepers and small business people protesting official efforts at taxation and price control. But, most characteristic is protest from casual workers and participants in the informal economy over the problems of collective consumption. These protests include land invasions, the demands of community organizations for land titles and public services, demonstrations and boycotts associated with the cost of transportation, and so forth. Because the informal economy lacks direct wage income, its protests are aimed at the indirect or social wage represented by urban services. Not only is this the most common new form of class struggle in the Third World city, it is a result that would be expected from a careful working through of the theory of dependent urbanism in light of major changes in the labor force.

Although based mainly on these issues, strategies and styles of protest (and negotiation) vary from clientism to autonomous action. Clientism is seen by some as cooptation (with reason), but more subtly by others as instrumental political "realism." Since the urban poor lack decisive political power and are vulnerable to shifts in public policy, they try to turn to their own advantage the manifestly cooptative efforts of government and charitable groups to promote "social mobilization." The urban poor are not naive about the bargain, but see it as a rare occasion to extract concrete concessions and to play one patron against another. The "severability" of these ties makes the poor less ready targets of repression during times of crisis (Leeds and Leeds, 1976).

Autonomous action is a more radical and risky alternative involving organizations of the urban poor outside of dominant institutional arrangements — organizations that aggressively press for more sweeping solutions such as legalized communal ownership of urban land or a bill of "urban social rights." Advocates of this open form of class struggle argue that it is generally more successful (avoiding virtually certain manipulation, division, and cooptation entrained with clientism), although it is undeniably dangerous and depends on broader alliances with peasants and the entire (formal sector) working class (Castells, 1976). The effectiveness of these two strategies is a complex and unsettled question that probably depends upon the concrete circumstances of political force and resource (Walton, 1979). For the moment, however, both are common, typically combine, and illustrate the major contours of urban class struggle.

CONSEQUENCES

Assuming that I have made a better case for the two amended assumptions and their theoretical implications than for the stark mechanics of the theory of peripheral urbanization, the final question concerns the consequences of this view. What can we expect from the new exigencies, social movements, and state and class structures of the Third World city? A general answer (and the only appropriate one here) is that two new realities govern the situation, one characterizing the institutional system and another the community. From the institutional standpoint, the sheer problem of accommodating the urban poor colors everything else: "Provision of shelter and urban services by the State to a large proportion of a deprived urban population is one of the major channels for political participation and community organization in the new institutional system emerging from a revival of national states struggling for control in a context provided by an international economy" (Castells, 1980:93). This succinct statement suggests, like the previous discussion, that three institutional foci become more salient and critical in their interaction, namely organized communities of the urban poor, "revived" (nationalistic) state structures, and the community of international interests. The negotiated interplay of these three may well produce a result different from the imagined acquiesence of "peripheral states" under the domination of core capital.

The new reality of the urban community is that "the quest for a new secure space is a major step in the search for preserving cultural identity, improving their living conditions, and ensuring political self-determination. *The growing urban population of the Third World is clearly oriented toward the building and preservation of spatially defined local communities*" (Castells, 1980: 94). Again, the terse statement points to considerations previously emphasized such as the cultural significance of urban communities and their recourse to political strategies for ensuring a permanent place in the city's economy and social life.

In summary, the unobjectionable idea of urbanization under peripheral capitalism should convey a great deal more than domination from afar. The emerging city of the Third World shows us a complex mixture of distant and local influences that defy additive or linear interpretation. At the bottom the consequences of these interacting influences are decided by people organized in classes and communities. Their organizations compete with others on state terrain and according to rules of a game that change as much as a result of shifting state ambitions as from their own impress.

It is not the proper task of social theory to oversee these changes, or selected portions of them, to sum them up, and to judge the "net result" (itself an illusion) in terms of optimism and pessimism — or to retreat into self-centered ennui. That is bad moralizing and bad theory. On the contrary, the job of research and theory is to capture the main effects *and* their *variation* — the *conditions* under which various *alternatives* are apt to flourish. Good theory is always optimistic.

REFERENCES

AMIN, S. (1974) Accumulation on a World Scale: A Critique of the Theory of Underdevelopment. New York: Monthly Review Press.

BALAN, J., H.L. BROWNING, and E. JELIN (1973) Men in a Developing Society: Geographic and Social Mobility in Monterrey, Mexico. Austin: Institute of Latin American Studies, University of Texas Press.

BERRY, J.L. (1964) "City size distributions and economic development," pp. 138-152 in John Friedmann and William Alonso (eds.) Regional Development and Planning: A Reader. Cambridge, MA: MIT Press.

BIRKBECK, C. (1979) "Garbage, industry and the 'vultures' of Cali, Colombia," pp. 161-183 in Ray Bromley and Chris Gerry (eds.) Casual Work and Poverty in Third World Cities. New York: John Wiley.

BROMLEY, R. and C. GERRY (1979) Casual Work and Poverty in Third World Cities. New York: John Wiley.

CARDOSO, F. H. and E. Faletto (1969) Dependencia y Desarrollo en América Latina: Ensayo de Interpretación Sociólogica. Mexico: Siglo Vientiuno Edition.

CASTELLS, M. (1980) "Multinational capital, national states, and local communities." Research Monograph Series. Tokyo: United Nations University.

——— (1978) The Urban Question: A Marxist Approach. Cambridge, MA: MIT Press.

——— (1976) "Movimentos socials urbanos en América Latina: tendéncias históricos y problemas teóricas." Madison: University of Wisconsin Press. (mimeo)

——— (1972) Imperialismo y Urbanización en América Latino. Barcelona: Editorial Gustavo Gili.

CORNIELIUS, W. A. (1975) Politics and the Migrant Poor in Mexico City. Stanford, CA: Stanford University Press.

FRANK, A. G. (1967) Capitalism and Underdevelopment in Latin America: Historical Studies of Chile and Brazil. New York: Monthly Review Press.

GORDON, D. (1978) "Capitalist development and the history of American cities," pp. 25-63 in William K. Tabb and Larry Sawers (eds.) Marxism and the Metropolis: New Perspectives in Urban Political Economy. New York: Oxford University Press.

GUGLER, J. and W. G. FLANAGAN (1978) Urbanization and Social Change in West Africa. Cambridge: Cambridge University Press.

HART, K. (1973) "Informal income opportunities and urban employment in Ghana." Journal of Modern African Studies 11: 61-89.

HARVEY, D. (1973) Social Justice and the City. Baltimore: Johns Hopkins University Press.

KENTOR, J. (1981) "Structural determinants of peripheral urbanization: the effects of international dependence." American Sociological Review 46 (April): 201-211.

LAQUIAN, A. A. (1971) Slums are for People: The Bario Magsaysay Pilot Project in Philippine Urban Community Development. Honolulu: East-West Center Press.

LEEDS, A. and E. LEEDS (1976) "Accounting for behavioral differences: three political systems and the responses of squatters in Brazil, Peru and Chile," pp. 193-248 in John Walton and Louis H. Masotti (eds.) The City in Comparative Perspective: Cross-National Research and New Directions in Theory. Beverly Hills: Sage.

LEWIS, O. (1952) "Urbanization without breakdown: a case study." Scientific Monthly 75 (July).

LOJKINE, J. (1976) "Contribution to a Marxist Theory of capitalist urbanization," pp. 119-146 in C. G. Pickvance (ed.) Urban Sociology: Critical Essays. New York: St. Martin's.

McGEE, T. G. (1979) "The poverty syndrome: making out in the Southeast Asian city," pp. 45-68 in Ray Bromley and Chris Gerry (eds.) Casual Work and Poverty in Third World Cities. New York: John Wiley.

——— (1971) The Urbanization Process in the Third World. London: C. Bell.

McGreevey, W. P. (1971) "A statistical analysis of hegemony and log-normality in the distribution of city size in Latin America," in Richard N. Morse (ed.) The Urban Development of Latin America, 1750-1920. Stanford, CA: Center for Latin American Studies.

MOORE, B., Jr. (1966) The Social Origins of Dictatorship and Democracy: Lord and Peasant in the Making of the Modern World. Boston: Beacon Press.

NERUDA, P. (1976) Memoirs. Trans. Hardie St. Martin. New York: Farrar, Straus, & Giroux.

OXALL, I., T. BARNETT, and D. BOOTH (1975) Beyond the Sociology of Development: Economy and Society in Latin America and Africa. London: Routledge & Kegan Paul.

PERLMAN, J. (1976) The Myth of Marginality: Urban Poverty and Politics in Rio de Janeiro. Berkeley: University of California Press.

PICKVANCE, C. G. (1976) Urban Sociology: Critical Essays. New York: St. Martin's Press.

PORTES, A. and J. WALTON (1981) Labor, Class, and the International System. New York: Academic Press.

——— (1976) Urban Latin America: The Political Condition from Above and Below. Austin: University of Texas Press.

ROBERTS, B. R. (1978) Cities of Peasants: The Political Economy of Urbanization in the Third World. London: Edward Arnold.

SLATER, D. (1979) "The state and territorial centralization: Peru, 1968-1978." Boletín de Estudios Latinoamericanos y del Caribe 27 (December): 43-67.

——— (1978) "Towards a political economy of urbanization in peripheral capitalist societies: problems of theory and method with illustrations from Latin America." International Journal of Urban and Regional Research 5 (March): 26-52.

——— (1975) "Colonialism and the spatial structure of underdevelopment: outlines of an alternative approach, with special reference to Tanzania." Progress and Planning 5, 4: 137-162.

THOMPSON, E. P. (1963) The Making of the English Working Class. New York: Vintage.

WALLERSTEIN, I. (1980) The Modern World System II: Mercantilism and the Consolidation of the European World Economy, 1600-1750. New York: Academic Press.

——— (1974) The Modern World System: Capitalist Agriculture and the Origins of the European World Economy in the Sixteenth Century. New York: Academic Press.

WALTON, J. (1981a) "The new urban sociology." International Social Science Journal 33, 2: 374-390.

——— (1981b) "Comparative urban research." International Journal of Comparative Sociology 22 (March-June).

——— (1979) "Urban political movements and revolutionary change in the third world." Urban Affairs Quarterly 15 (September): 3-21.

——— (1976) "Guadalajara: creating the divided city," pp. 25-50 in W. A. Cornelius and R. V. Kemper (eds.) Metropolitan Latin America: the Challenge and the Response. Latin American Urban Research. Vol. 6. Beverly Hills: Sage.

7

Regional Development and National Integration:
The Third World

SALAH EL-SHAKHS

□ A NEW FOCUS has recently been emerging in both regional development theory and policy as a culmination of a consistent shift, over the past twenty-five years, from a "technical" to a "sociopolitical" emphasis (Rodwin, 1978). Such a shift has been a reaction to the largely disappointing results of past regional development strategy experiences in Third World countries, the minimal and often adverse impact of national economic development gains on the lot of disadvantaged regions and the poor in general (Gilbert, 1978; Hauser, 1979), and better insights into the political nature of the spatial development process.

It is not uncommon that the forces created by national integration and development policies are in conflict with the objectives promulgated by regional development strategies (Renaud, 1979; Stöhr, 1975). This chapter reviews our current knowledge of spatial development theory and experience in relation to such conflicts and their implications for regional development strategies. It suggests that legitimate efforts of national integration in market and mixed economy systems of the less developed countries (LDCs) tend to increase regional disparities. It further recommends some regional development approaches and critical concerns worthy of concerted future research efforts in the field.

AUTHOR'S NOTE: This chapter is based in part on a study prepared by the author for a United Nations Fund for Population Activities' conference on "Population and the Urban Future" held in Rome in September 1980.

REGIONAL DEVELOPMENT POLICY IN RETROSPECT

National interest in explicit regional development strategies in LDCs is scarcely three decades old. The prevailing wisdom then was that local comprehensive land use and infrastructure development plans coupled with vigorous national economic growth were all that was needed to improve economic and social conditions uniformly throughout a spatial system. It was soon recognized that the spread effects of economic growth were not as strong as expected, and development did not trickle down to less advantaged regions. It was felt that such regions needed to be targeted for specific concentrated doses of propulsive-type investments in regional development centers (growth centers). The basic premise of such policies was that development will not only intensify at such strategic nodes, but will also filter down over time, spreading income and employment throughout their regions. Rejecting such assumptions is one of the few points on which there seems to be a consensus among regional development specialists both from the left as well as the right (Rodwin, 1978). It is argued, albeit for different reasons, that the critical urban and economic linkages do not favor the disadvantaged hinterlands, and that the resulting backwash effects tend to overpower any spread effects.

The focus of regional development theory, if not practice, thus shifted to the articulation of sociopolitical forces and the nature of such linkages at both the national and international levels. Rodwin (1978) provides an excellent overview of the varying interpretations of such forces which range from the exploitative nature of the capitalist system (Sunkel, 1969; Slater, 1975 and 1977; Amin, 1976; Frank, 1967 and 1970 among others), to the inherent dualism favoring the modern sector and its elite (Weeks, 1975; McGee, 1976, among others), to blaming the inefficiencies and lags built into the modernization process. A common underlying theme in these views is the conviction that regional development is a function of the process of national and international economic and political integration.

The recent *World Development Report 1980* (World Bank, 1980: 1) points out that the LDCs face two major challenges during the coming decade:

> First, they must strive to continue their social and economic development in an international climate that looks less helpful than it did a decade — or even a year ago. Second, they must tackle the plight of the 800 million people living in absolute poverty, who have benefited much too little from past progress.

Indeed the lot of the poor has worsened — in real terms — despite impressive national urbanization and economic gains in several LDCs during the 1970s. Aggregate growth in either respect has been marred by distributional imbalances which favored concentration in supercities or core metropolitan areas. Current United Nations projections expect such trends to continue (see Table 7.1).

The challenges to national regional development planning in the LDCs thus lie primarily in its ability simultaneously to expand the economy and equitably distribute its resources in an effort to hasten the spatial transition process. In a recent United Nations conference on Population and The Urban Future (UNFPA, 1980: 3) the assembled policy makers and planners predictably concluded that:

> To improve and enrich the quality of life for (the World's) increasing numbers of urban dwellers, . . . a fairer distribution of wealth among nations is necessary. At the same time internal changes are necessary to ensure an equitable distribution of resources and a fair and just society within each nation. The objective must be to achieve a balanced allocation of resources and development opportunities and of the economic and social benefits resulting from them.

Such declaration underscores the perceptible shift in regional development policy:

(1) from strong faith in the spread effects of overall economic growth (efficiency) to a recognition of the need for explicit redistribution strategies (social and spatial equity);

(2) from focusing on an individual region's (or country's) internal inadequacies, inherent qualities, and development potential to stressing its position and linkages as an integral part of a national and international economic and political system; and

(3) from primary emphasis on economic dimensions and factors in regional development to an attention to the psychosocial environment, regional identity, and political power.

The reasons for such shifts are to be found in the generally disappointing experience with regional development strategies in market and mixed economies, where market forces are allowed to influence the distribution of urbanization and economic functions through freedom of movement of population and factors of production. This is largely due to the inability (and/or unwillingness) of most countries to initiate and implement integrated development policies (Renaud, 1979), and an inadequate understanding of the dynamics of spatial development processes (El-Shakhs, 1980).

TABLE 7.1   World Population Projections by Rural, Urban, Large City and Super City for MDCs and LDCs 1980-2000

		1980		2000		Percentage Expected Increase 1980-2000
		Number in Billions	Percentage of Total	Number in Billions	Percentage of Total	
Super Cities (1,000,000+)	LDC	0.34	8	0.93	15	173
	MDC	0.31	7	0.43	7	39
	Total	0.65	15	1.36	22	109
Large Cities (100,000+)	LDC	0.62	14	1.49	24	140
	MDC	0.55	13	0.75	12	36
	Total	1.18	27	2.24	36	89
Urban Population	LDC	0.97	22	2.12	34	118
	MDC	0.83	19	1.09	17	31
	Total	1.80	41	3.21	51	78
Rural Population	LDC	2.21	41	2.75	44	24
	MDC	0.35	8	0.29	5	−17
	Total	2.56	59	3.04	49	18
Total Population	LDC	3.18	73	4.87	78	53
	MDC	1.18	27	1.38	22	17
	Total	4.36	100	6.25	100	43

SOURCE: United Nations, Patterns of Urban & Rural Population Growth (New York: U.N., 1980).

The integration of policy, in terms of economic and social objectives, sectoral and spatial implications, and short-term and long-range impacts, demands a clear perspective on the forces of spatial development, their timing, and the conditions which trigger them. Its effectiveness requires a high degree of commitment on the part of policy makers and planners to articulate, coordinate, and carry out such policies. Economic plans tend to be spatially biased, and spatial development strategies tend to underestimate the power of economic linkages and psychosocial forces, or ignore their own economic and institutional limitations (Stöhr, 1975; Renaud, 1979). Integrating economic and spatial development plans, therefore, is a primary planning issue which requires a better understanding of the consequences, limitations, and trade-offs involved in such plans.

Experience indicates that well-intentioned efforts of spatial and economic decentralization, such as expanding transportation networks and decentralizing modern industries, while designed to reduce development inequalities or urban concentration or both, tend to produce opposite results. They generally tend to unleash powerful polarization forces which overpower embryonic regional economies and hierarchical settlement patterns, focus attention and aspirations on large cities, and enforce national and regional urban concentration and primacy, particularly in small countries where space does not present major accessibility barriers (Stöhr and Tödtling, 1978).

Thus, the inadequacy of our knowledge of the dynamics of spatial development processes severely limits our ability to predict their long-range patterns and the capacity to shape them through public policy. This is particularly true in rapidly changing LDCs with market and mixed economies, where spatial development patterns are characterized by major time-lags and may be discontinuous over the long range. Spatial development strategies based on uncertain long-range predictions could become restrictive rather than adaptive and prove to be costly in the long run.

THE PROCESS OF SPATIAL DEVELOPMENT

Current descriptive theory of national spatial development took on a new momentum with the dissemination of two major findings in 1965 concerning the spatial dynamics of the development process. The first identified a bell-shaped curve of regional income inequalities with development (Williamson, 1965). The second simultaneously identified a similar curve for urban concentration (or degree of primacy within urban systems) with development, and suggested that similar curves would also apply to the rates of aggregate national economic growth, urbanization, and population growth (El-Shakhs, 1965, 1972). The outlines of a general theory of spatial development therefore implied a succession of two broad trends: first, toward the geographic concentration of urban population, economic activities, and income during the early stages of national development; then, second, a reverse process of deconcentration or dispersion, both intraregional and interregional, as nations reached mature levels of economic and social development.

Clearly such a broad long-range perspective on the spatial development process needed to be explained in terms of a wide range of relevant variables within diverse geographical and cultural contexts. Furthermore, since empirical observations relied heavily on the experience of those advanced nations which have undergone the full range of the process, it has been argued that such historical experience of more developed countries (MDCs) is of little relevance to the future of LDCs because of major differences in internal developmental contexts and their respective roles in the international system.

Such criticisms notwithstanding, the first part of a general theory of polarized development (Friedmann, 1973), explained in terms of unbalanced growth and cumulative causation effects favoring one or a few advantaged core regions, is currently far less disputed, in either theory or experience, than the subsequent process of polarization reversal marked with interregional convergence in levels of development and welfare (Richardson, 1977 and 1980). Indeed the latter part of this process (convergence) has attracted far less attention and explanation in the regional development literature (Alonso, 1980), partly because empirical evidence is relatively limited, and partly because that which exists relates to advanced nations and did not until very recently present urgent issues of national or international public concern.

The process and issues of intrametropolitan or intraregional dispersion — or what Berry (1976) termed the process of "counter-urbanization" — commanded more attention in the United States and in some countries in Western Europe. Within the broader context of spatial development, however, this process is not synonymous with that of interregional deconcentration, though its occurrence, at least in the primary core regions, may precede and signal the onset of interregional polarization reversal. Indeed, as recently as the 1960s, the United States and several Organization of Economic Cooperation and Development (OECD) countries were still preparing for major projected growth in their largest core metropolitan regions with concern only for problems of their internal dispersion. Currently, however, they are busy trying to cope with the emerging decline of these areas as a consequence of major interregional shifts of population and economic activities. A recent study (Vining and Kontuly, 1978) indicated that several of these countries have recently been witnessing both relative and/or absolute declines in the populations of their major core regions. Indications are that several middle-income countries (e.g., the Republic of Korea, Chile, Taiwan, and Spain) may soon be approaching a similar process of polarization reversal.

Recent developments in the LDCs may also be instructive in this respect. While, given relatively low levels of urbanization, most of them may still have to cope with urban explosion over the coming generation (Table 7.1), it is clear that the corner on the population explosion "has turned much faster than any one predicted" (Abu-Lughod, 1977). Fertility rates have begun to decline in several LDCs, faster than they did in the more developed countries, while gross domestic products (GDPs) have generally been growing faster ⁺han total populations.

A recent analysis of the 33 countries with the world's largest urban agglomerations (El-Shakhs, 1980) also indicated that secondary and medium-size cities (100,000 to 1 million in population) registered substantial proportionate gains in population growth sometime during the last three decades in such countries as the Republic of Korea, Mexico, Egypt, and Brazil. The rate of increase in the proportion of urban population in cities over 1 million has been declining in these countries in addition to the Philippines, India, and Spain. In general, this rate is projected to decline drastically between 1980 and the year 2000 in all of the world's less developed regions (Hauser and Gardner, 1980). In addition, spatial deconcentration and expansion of large urban agglomerations (intraregional) are already taking place in most developing as well as developed nations.

Clearly such indications require additional time and empirical research to be confirmed as trends. Nevertheless, they do support the long-range spatial development pattern of interregional concentration — deconcentration with development.

Regions within national spatial systems are likely to follow a similar pattern, in which the most developed region (core area) is likely to be the first to undergo a process of intraregional deconcentration (or dispersion) followed by other regions as they reach appropriate concentration and economic and social development thresholds (Figure 7.1A).

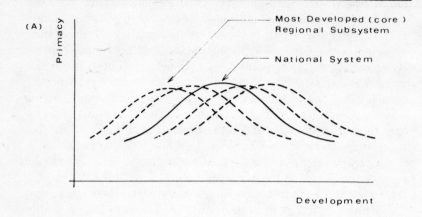

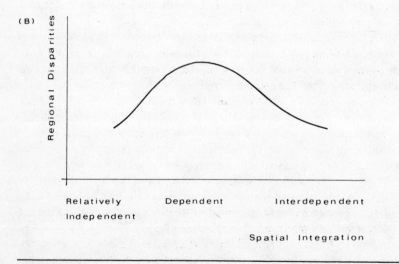

Figure 7.1 Patterns of Change in Primacy with Development and in Regional Inequalities with National Integration

SOURCE: El-Shakhs, 1980.

While such a long-range perspective on ultimate changes within the spatial system is useful as both a general guide and as a target for long-range planning, it provides few practical clues for the interim design of public policy. The timing and concentration thresholds of polarization reversal, the form of the resulting size and spatial hierar-

TABLE 7.2 Urban Concentration, Income Inequality, and Per Capita GDP In Selected Countries with Large Urban Agglomerations

Nations in Rank Order of Per Capita GDP (1976)	Urban Population as Percentage of Total (1980)	Population of Largest City as Percentage of Urban Population (1980)	Gross Domestic Product 1976 (Billion U.S. Dollars) National	Per Capita	Income Inequality Index Regional[1] (1976)	National[2]
Burma	27	35	3.5	113	—	.381
India	22	6	86.0	141	.185	.477
Pakistan	28	24	14.5	200	—	.330
Indonesia	20	22	37.3	267	—	.463
Thailand	14	48	16.3	379	.678	.448
Columbia	70	22	14.8	608	.307	.562
Korea, R. of	55	47	25.4	707	.308	.372
Mexico	67	30	79.1	1,270	.534	.583
Brazil	65	16	144.7	1,325	1.620	.647
Iraq	72	53	15.7	1,363	—	.629
Argentina	82	47	47.4	1,844	.313	.411
Iran	50	32	66.8	1,988	.923	.502
Venezuela	83	26	31.1	2,510	.533	.477
Spain	74	15	104.7	2,908	—	.393
Italy	69	17	170.8	3,040	.272	—
United Kingdom	91	25	219.2	3,922	.109	.339
Japan	78	21	555.2	4,922	.301	.287
France	78	23	346.6	6,552	.243	.518
Germany, F.R.	85	19	445.0	7,249	.135	.398
United States	73	10	1,701.7	7,911	.134	.407

SOURCES: El-Shakhs, 1980 — adapted from Tables 1 and 3; and World Bank, 1980, Table 20.

[1] The regional inequality index is the coefficient of variation. This is the ratio of the standard deviation of regional per capita incomes (weighted according to the relative population size of regions) to the mean regional per capita income. The index is directly related to the level of inequality among regions, decreasing to a minimum value of zero as total equality is achieved.

[2] This is the Gini coefficient. It is directly related to the level of national inequality, falling to zero as total aggregate national equality is achieved. For this sample, the Gini, Kuznets, and entropy coefficients of aggregate income inequality are highly correlated. All the Gini coefficients pertain to years since 1960.

chy, and the optimality of such distributions can vary considerably with the spatial, economic, and political development contexts of countries (Richardson, 1977; von Boventer, 1973).

The divergence in experiences will be closely related to differences in the initial conditions and relative distribution within the system, economic and political structure, and the nature of the processes of international and national integration.

SYSTEM INTEGRATION AND REGIONAL DISPARITIES

High degrees of concentration in spatial systems, therefore, appear to be an inevitable consequence of the cycle of political and economic polarization-dependence characteristics of the early stages of development under market conditions. This cycle is usually reinforced by unfavorable conditions in integration within the international economic system and the unequal basis of political and economic integration within national spatial systems, which tend to increase interregional disparities (Figure 7.1B). The latter is not only associated with increased population concentration in core regions (Table 7.2), but indeed often credited with causing them (Stöhr and Tödtling, 1978).

At the beginning of the process of national integration, in traditional societies, spatial units or regions are relatively self-sufficient and independent of each others' economic and political influence. Minimal levels of urbanization, the weak state of communications and spatial interaction, and the tribal or regional nature of authority and political power tend to act as barriers against strong hierarchical dependencies or dominance. However, initial disparities in development levels and potential are likely to exist among such spatial units as a result of their human and natural resource bases, locational advantages, and historical ties and interactions with the rest of the world (El-Shakhs, 1965).

Increased spatial interaction, urban industrial development, and economic integration among such units (regions or countries) tend to accentuate their initial disparities and reinforce development dependencies favoring those relatively advantaged units. This process has frequently been intensified by external or internal colonization policies, increased flows of labor and capital, economies of scale and of agglomeration, and a "cumulative causation" process of concentration. Spatial systems thus enter a process of dependent development and integration focused on their emerging core regions.

An exception to this process may exist if the local markets of the less developed regions were protected by their relative isolation until they developed sufficient scale economies (in terms of population and income) to generate or attract industries that cannot serve them from the core region. This may have been the case in the experience of the MDCs (Richardson, 1977). Also, relative isolation may explain the low levels of national primacy and inequalities in very large countries, where long distances, the large size of regional subsystems, and strong regional identities act as barriers against strong national spatial and functional accessibility. Such systems would tend to develop strong intraregional concentration and primacy patterns focused on large regional centers (e.g., India and China).

Other exceptions to the pattern of increasing disparities with integration, in the early stages of development, may exist as a result of strong measures of redistribution, including dispersed political power, strong controls of both flows and means of exchange (centrally planned countries), and/or specialized resource endowments that favor the less developed areas.

If countries reach higher levels of development, they seem to develop a strong negative relationship between national prosperity and regional inequality, i.e., higher levels of national prosperity signify greater equality among subnational regions. Greater integration at those levels (i.e., interdependent integration) would tend to speed the process of regional equality and thus increase national aggregate equality as well, since experience indicates a strong positive relationship between the two (El-Shakhs, 1980).

These patterns imply an eventual reversal in the relationship between the extent of national integration and regional disparities and sustained national economic growth. Indeed, declines in regional disparities in development levels and urban concentration have largely been attributed to autonomous changes and development within the national economies, rather than to the effects of public policy (Mera, 1978). This does not necessarily imply, however, that disparities could not be reduced through public policy; rather, that the policies may not have been appropriate, or may have been applied at the wrong time, within their economic and political contexts.

There is some evidence that policies designed to correct an imbalance within the system, during stages of polarization, serve instead to intensify that very imbalance or to create new demands, in the long run, which are more difficult or costly to satisfy (United Nations, 1977). Paradoxically, programs intended to speed the spread effects from core regions and to develop the hinterlands sufficiently to retain their populations, often result in increasing migration to these cores.

The eventual reversal of polarization trends seems to be contingent on achieving an equitable social and spatial distribution of development benefits and of political power. Such redistribution, however, could entail high economic costs, under conditions of underdevelopment, and/or high political costs, under conditions of ethnic and spatial diversity. On the other hand, unmitigated polarization processes could lead to similar costs, particularly in the long run.

Many governments have recognized that it would be both economically wasteful and politically and socially unacceptable, or intolerable, to wait patiently for an automatic polarization reversal process which would eventually significantly reduce regional disparities and the pressures on very large cities. Even if such acceptance of dependent integration and increased economic polarization, as an implicit or explicit development strategy, were to result in significant increases in GNP, experience of LDCs has shown, as pointed out earlier, that GNP growth often fails to improve the conditions of the poorest people in both rural and urban areas. This "growth without development" approach would also have adverse implications for the long-range economic efficiency of the system and the short-range functional and orderly development of the large cities (Slater, 1977).

A recent survey indicated that 83 percent of the LDCs (92 percent in Latin America and 94 percent in Africa) perceive the spatial distribution of their population as unacceptable (United Nations, 1979). Dissatisfaction with primate city patterns and recognition of the need to rationalize population-resource relationships focused attention on issues of regional development and the structure of the urban system, and the resulting spatial and hierarchical imbalances. As a result many governments have attempted to influence interregional development patterns through indirect national policies or direct spatial development strategies, and nearly half of the LDCs formulated policies to modify the distribution of their urban population.

POLICY RESPONSES

Redistribution policy and strategy responses can generally be classified, in terms of their focus and basic assumptions, into two categories:

(1) manipulation, regulation and control of the *economic basis* of development through transfer and distributional allocations and instruments, legal and administrative controls of factor movements, and spatial integration and diffusion process; and

(2) creation of the *social, psychological, and political environment* for development through political and administrative reorganization, redistribution of decision-making power and control over development, initiative and participation, and local integration and self-reliance approaches.

The first seems to have figured more prominently in most experiences to date (e.g., growth centers, industrial decentralization, rural development, transport expansion and integration with core areas, and so on). Their success, particularly during stages of concentration, seems to have at best been very limited. Such integration efforts are based on neoclassical assumptions about flows of labor and capital, and assumptions that innovations, and therefore development and income, will diffuse down the urban hierarchy and spread out of urban centers into their hinterlands. It appears, however, that economic decentralization and diffusion strategies tend to be relatively more successful where spatial inequalities are low, local governments are strong and relatively autonomous, and local resource endowments are high (Stöhr and Tödtling, 1978).

In the absence of tight control of different aspects of society (centrally planned economies) or existence of alternative outlets in the form of greater options of choice (MDCs), direct administrative controls are likely to be ineffective. Closed city programs that seek to restrict migration to large cities (Soviet Union, Poland, China, Indonesia, Italy), reduce or restrict additional investment (France, Egypt), limit the growth of the informal sector (through harassment), deport illegal residents (the United Republic of Tanzania, the United States), or deny them services (Philippines) have only been partly successful at best (Findley, 1977). Efforts to remove squatters and nonlicensed activities or to control the use of urban and agricultural land often meet with corruption and/or political pressure. Furthermore, planning efforts have frequently tended to emphasize negative controls rather than positive adaptive approaches. This makes their implementation unlikely to be effective, even if the necessary administrative coordination, persistence, and resources are available.

On the other hand, the positive effects of intergovernmental fiscal relations (Japan, the Republic of Korea), regionalization of the budget (France, Sweden) and revenue sharing with a measure of local autonomy (Nigeria, the United States) indicate the importance of fiscal and administrative decentralization as a redistribution strategy. In an economic sense, it is a redistribution of public and government employment, which, in most LDCs and centrally planned economies, accounts for a substantial proportion of total formal sector employment, personal income, and purchasing power. Institutionally, it provides an important component of the social infrastructure and local organizational capacity, and it nurtures the creation of the demand for, and supply of, local services and amenities. Politically, its effect could range from the simple creation of a local bureaucratic constituency (an interest group within the national bureaucracy), to broadening the base of decision-making, participation, and initiative (redistribution and devolution of power).

It is therefore important to focus our attention on the effects of redistribution of effective power, broadening the base of decision-making, and local initiative and participation. The psychological, social, and political factors of development need to be better understood within the context of integrated functional-area planning at local and regional levels. The division of authority and responsibility among different levels of government, including those for planning and budgeting, raises important issues. Frequently, municipal authorities are delegations of national government power and depend on national decisions for the major part of their financing, organization, and operation. This often results in weak local governments that lack the power, resources, and flexibility to respond to changes within their own boundaries. It may also be nonconducive to the development of strong leadership or a clear local identity.

REGIONAL POLICY IMPLICATIONS

Clearly, it is a legitimate and primary function of national governments to promote national integration within their social and spatial systems. This is particularly true for those LDCs with diverse regional and ethnic constituencies and colonial legacies, where national identity and territorial integrity become overriding concerns above all other development goals. However, the impact of national integration strategies on development disparities (interregional, interurban, and interpersonal) depends to a large extent, as pointed out earlier, on the level of initial disparities in resources, levels of development, and effective shares in political power.

In the absence of effective balancing redistributive measures, therefore, the process of integration may lead to greater divergence within the national system. By pursuing national integration goals, governments cannot be neutral in such a process. Furthermore greater integration results in increased information flows and exchange which necessarily lead to stronger perceptions and awareness of imbalances and disparities (Stöhr and Tödtling, 1978). The resulting inflation of expectations and heightening of conflicts increasingly become primary issues of national public policy. Paradoxically, such consequences of national integration efforts could become, under certain conditions, a threat to the cherished goal of maintaining national territorial integrity (Richardson, 1977).

The policy issue, therefore is not whether national governments should intervene in the spatial development process, by actively promoting national integration, but what type of role should they play, when, and how? (Stöhr, 1979). Clearly, traditional market forces do not operate, at least in the short run and particularly in LDCs, to create the equilibrating and optimizing mechanism attributed to them by neoclassical economics (Hansen, 1978). Thus, even if long-term economic and spatial convergence is an ultimate eventuality in the development process, there are compelling demographic, social, and political reasons to speed it up through public policy.

If properly timed and designed, government intervention should enhance rather than retard balancing mechanisms. It should reduce the time lag in interregional and interurban equilibrium by moderating the differences in access to urbanization and scale economies. It should widen rather than restrict the freedom of choice of potential migrants by increasing viable alternative options, and their knowledge of them, including that of not migrating. A cornerstone of such efforts is to strengthen rather than weaken the local governments' identity and ability to respond to local challenges and to manage and control their local resources effectively.

The problems with such an approach can be summed up in five basic arguments. First, polarization (or more specifically a certain level of spatial agglomeration and concentration of economic activities) may be necessary for economic growth (Alonso, 1968; Mera, 1973). Second, balanced development policies may be costly and ineffective during the period when polarization forces are strong (Richardson, 1980). Third, polarization reversal and subsequent convergence result primarily from response to autonomous forces rather than from the impact of public policy (Mera, 1978). Fourth, politically and institutionally, LDCs are unwilling or incapable of initiating and coordinating integrated policies. Finally, premature regional convergence may preempt the development of effective national political and governmental institutions and encourage forces of disintegration (Salau, 1980).

Such arguments certainly merit our attention and point to directions for needed further research since equally plausible counterarguments can also be made. Continued economic growth seems to be a necessary but not a sufficient condition for reducing spatial and social inequalities. While a certain degree of concentration and therefore disparity may be necessary to achieve adequate economies of agglomeration and scale, if the gross national product is to grow rapidly during early stages of development, the notion of optimality or the need for continued concentration beyond minimum thresholds is much more questionable. Optimality is a very subjective concept related to level of development and development goals, capacity to plan, and culturally determined tolerance levels of interaction.

If the rapid growth of the largest city is not associated with, or necessary for, major additional industrial expansion within the system, and if the capacity to organize, plan, and service very large cities is low (which is the case in most LDCs), then pursuing hierarchical balance and distributional equities, beyond the necessary minimum concentration thresholds, becomes the significant challenge for both short- and long-range policies. Regional development and integration policies should thus aim at containing such disparities within tolerable limits. The need to trade off efficiency for equity can be minimized by broadening the natural resource base for development planning, and strengthening the forces of interregional horizontal spatial expansion and integration (Stöhr, 1975).

In their pursuit of national integration and economic growth, policy makers in the LDCs should become more cognizant of the unintended consequences of their policies on spatial and social inequalities. The issue, during concentration stages, is to avoid reinforcing polarization trends and greater disparities rather than attempting to reverse them. In addition to their potential threat to the very goal of national integration, polarization-reinforcing policies tend to result in the concentration of costly "space-forming" urban industrial infrastructure which would inhibit eventual polarization reversal, or at best make it more difficult and extremely wasteful in terms of natural, capital, and human resources. Policies should anticipate future spatial structures and, in the process of attempting to solve current problems, lay down the basis (in terms of both physical and institutional infrastructure) for their eventual emergence.

Since spatial development patterns and objectives, as well as government policies and commitments, are constantly changing, it is extremely difficult to distinguish between autonomous and policy-induced changes (Stöhr and Tödtling, 1978; Richardson, 1980). Where explicit regional development strategies are ineffective in inducing balanced development, they may be proven to have been inappropriate, applied at the wrong time in the spatial development process, or contradicted by strong implicit impacts of national policy and institutional structure. These cannot be neutral in shaping autonomous processes.

Ineffective strategies have by and large focused primarily on economic factors. Not only are the social, psychological, and political forces as important in shaping development, if not more so, but it is virtually impossible to base development strategies purely on economic engineering (Corragio, 1975). Effective regional development, therefore, requires the building of local political and governmental structures capable of responding to local challenges, internalizing the benefits of development, and counteracting excessive concentration trends.

This is perhaps the most critical unresolved issue of regional development and national integration. It requires an awareness on the part of national governments and core region elites of the comparative risks and inefficiencies involved in continued polarization compared to those of greater local fiscal and decision-making autonomy. Depending on the nature of the country's social and political structure, the choice could be between potential revolution and peaceful transformation. The latter, however, requires a clear articulation of the division of powers and authority, coordination among national and local governments, training and technical assistance, and a multilevel system of planning (Utria, 1972).

Facing the challenges of regional development in LDCs will require some hard political and economic choices on the part of both the LDCs and the community of nations at the international level. The focus of spatial development policy should no longer be on maintaining equilibrium or the status quo, but rather on promoting balanced interdependence. In terms of national policy, this suggests several points:

(1) More attention and priority should be given to efforts whose aim is to develop human resources, in a broader sense, and to improve the quality of life and meet basic human needs.

(2) Spatial development objectives should become an integral part of national economic and social development policies and plans.

(3) Efforts of national integration should enhance and broaden the base of local and regional decision-making power, administrative efficiency, and community participation and initiative.

(4) Development planning efforts should be articulated in terms of functionally integrated, relatively autonomous (economically and administratively) spatial subsystems at both regional and local levels, including integrated rural subsystems.

(5) Central governments should play an explicit balancing and facilitating role in the processes of spatial transition and regional adjustments through development and communication of basic information on population and economic activities, organization and education for planning, technical and fiscal support, and regionalization of planning and budgeting.

(6) Demographic, social, and psychological variables and objectives of population development should become integral parts of the planning and implementation process.

Spatial expansion and integration of the national economy require an expanded and efficient urban settlement system. The development of such a system, during the early stages of growth, particularly in large countries, requires conscious efforts and policies aimed at strengthening the economic, institutional, and psychological basis for local and regional development. Fostering local integration and identity, and creating a cultural milieu that enhances local pride and creative power, would help counter the polarization forces of national integration within the urban system.

One of the most sensitive and delicate policy choices, particularly for large, densely populated and diverse countries, is the need to balance the regionalization of development with the promotion of national integration. Such balance is particularly crucial in market and mixed economies, where spatial development can be *guided* through promoting options for, and influencing, individual choices, in contrast to *controlling* it through limited choices and planned inevitabilities in centrally planned economies.

Direct controls and restrictions of population movement, despite their relative success in a few instances, may have undesirable long-range demographic and economic consequences both for the largest cities and for the rest of the spatial system. Furthermore, they are difficult to enforce in most cases, particularly with increased accessibility and efforts of national integration in LDCs.

What is required in these cases, therefore, seems to be the initiation of a step-wise process of integration in which:

(1) Efforts to improve *intraregional accessibility* within and among peripheral regions and to help promote local regional integration and increase aggregate demand would take precedence over interregional and national spatial and economic integration efforts.

(2) The enrichment of social and cultural *amenities and basic services* in small urban places and regional centers, primarily through the redistribution of government and public sector

facilities and employment opportunities among regions and urban settlements, would take precedence over redistribution of modern industrial activities.

(3) Industrial development strategies would promote *competitive specialization* among subnational and regional centers on the basis of their locational advantages or regional resource base, rather than cheap labor.

(4) Administrative decentralization would genuinely strengthen the *redistribution of decision-making power* and control over resources, budget allocations, and development efforts to strengthen local identity, initiative, and participation.

A number of spatial development strategies have recently been advocated as a means of moderating the adverse impacts of national integration efforts on rural-urban and interregional disparities in LDCs along the above-mentioned principles. These include "integrated rural development" (Johnson, 1970), "agropolitan development" (Friedmann and Douglas, 1976; Lo and Salih, 1979), "selective spatial closure" (Stöhr and Palme, 1977; Stöhr and Tödtling, 1978). Such strategies aim primarily at generating, strengthening, and even protecting local development initiative within integrated rural-urban subsystems.

While local development initiative, with indirect national government support is crucial, it should be coupled with national counter-primacy efforts aimed at narrowing the urbanization inequities within the urban settlement system. This is particularly necessary in small countries and in countries which already have high degrees of urban concentration. A step-wise deconcentration strategy (interregional and subsequently intraregional) should focus on already established and autonomously growing (Richardson, 1977) specialized and/or regional service urban centers. The success of this strategy, however, would require the adoption of similar development principles emphasizing regional autonomy, social and physical infrastructure, income redistribution policies, local participation and initiative, and general improvement of the quality of life and business environment.

Such improvements would enhance the chances for subsequent decentralization of internal-market-oriented industries. Decentralization of export-oriented capital intensive industries, which are closely linked to the core region, should initially be avoided in order to minimize backwash effects and economic inefficiencies. These industries may be more appropriately decentralized within the core region subsystem (Richardson, 1978) as a prelude to their own eventual deconcentration and horizontal expansion into polycentric urban regions.

In conclusion, it is clear that spatial inequalities in welfare contribute to, and are influenced by, concentration and deconcentration trends. In effect, equalizing levels of welfare and quality of life among different regions and urban centers would not only contribute to moderating the impacts of such processes but also help eliminate the major cause for many local development problems. In mature urban systems, deconcentration and increased mobility also lead to inter-urban inequalities, at least in the short run, as a result of differential mobility rates and shifts in economic base.

In both cases, therefore, national and state governments should assume, either directly or through strong redistributive measures, responsibility for short-term satisfaction of a basic minimum level of social welfare across regional and municipal boundaries, and long-range equalization of such levels.

Similar goals should guide the specific objectives of international aid, and technical assistance programs of international agencies in dealing with priorities and channeling of development efforts both among and within nations.

REFERENCES

ABU-LUGHOD, J. (1977) "The urban future: a necessary nightmare?" Presented at a Rutgers University Distinguished Lecture Series, New Brunswick, N.J. To be published in J. Luitz and S. El-Shakhs (eds.) Tradition and Modernity. Washington, DC: University Press of America (forthcoming).

ALONSO, W. (1980) "Five bell shapes in development." Regional Science Association Papers 45: 5-16.

——— (1968) "Urban and regional imbalances in economic development. Economic Development and Cultural Change 17: 1-14.

AMIN, S. (1976) Unequal Development. New York: Monthly Review Press.

BERRY, B.J.L. (1976) "On urbanization and counter-urbanization," pp. 7-14 in B.J.L. Berry, Urbanization and Counter-Urbanization. Beverly Hills: Sage.

von BÖVENTER, E. (1973) "City size systems: theoretical issues, empirical regularities and planning guides." Urban Studies 10: 145-162.

CORRAGIO, J.L. (1975) "Polarization, development, and integration," pp. 353-374 in A.R. Kuklinski (ed.) Regional Development and Planning. Leiden: Sijthoff.

EL-SHAKHS, S. (1980) "National and regional issues and policies in facing the challenges of the urban future." Presented at the United Nations Fund for Population Activities conference on Population and the Urban Future, Rome, September.

——— (1965) "Development, primacy and the structure of cities." Ph.D. dissertation, Harvard University.

—— (1972) "Development, primacy and systems of cities." Journal of Developing Areas 7: 11-36.

FINDLEY, S. (1977) Planning for International Migration: A Review of Issues and Policies in Developing Countries. Washington, DC: Government Printing Office.

FRANK, A. G. (1970) Latin America: Underdevelopment or Revolution. New York: Monthly Review Press.

—— (1967) "Sociology of development and underdevelopment." Catalyst, 3: 20-73

FRIEDMANN, J. (1973) "A theory of polarized development," pp. 41-64 in J. Friedmann, Urbanization, Planning, and National Development. Beverly Hills: Sage.

—— and M. DOUGLASS (1976) "Regional planning and development: the agropolitan approach." pp. 333-387 in UNCRD, Growth Pole Strategy and Regional Development Planning in Asia. Nagoya.

GILBERT, A. G. (1978) "The dynamics of human settlement systems in less developed countries: the priorities for urban policy formulation," pp. 177-194 in N. Hansen (ed.) Human Settlement Systems. Cambridge, MA: Ballinger.

HANSEN, N. M. (1978) "Preliminary overview," pp. 1-21 in N. M. Hansen, Human Settlement Systems. Cambridge, MA: Ballinger.

HAUSER, P. (1979) World Population and Development, Ithaca: Syracuse University Press.

—— and R.W. GARDNER (1980) "Urban Future: Trends and Prospects." Presented at the United Nations Fund for Population Activities conference on Population and the Urban Future, Rome, September.

JOHNSON, E. A.J. (1970) The Organization of Space in Developing Nations. Cambridge, MA: Harvard University Press.

LO, F.-C. and K. SALIH (1981) "Growth poles, agrapolitan development, and polarization reversal: the debate and search for alternatives," pp. 123-154 in W. Stohr and D.R.F. Taylor (eds.) Development from Above or Below? New York: John Wiley.

McGEE, T. G. (1976) "The persistence of the proto-proletariat: occupational structures and planning of the future of third world cities." Progress in Geography, 9: 1-38.

MERA, K. (1973) "On the urban agglomeration and economic efficiency." Economic Development and Cultural Change 21: 309-324.

—— (1978) "Population concentration and regional income disparities: a comparative analysis of Japan and Korea," pp. 155-75 in N. Hansen (ed.) Human Settlement Systems. Cambridge, MA: Ballinger.

RENAUD, B. (1979) National Urbanization Policies in Developing Countries. World Bank Staff Working Paper 347. Washington, DC: World Bank.

RICHARDSON, H. (1980) "Polarization reversal in developing countries." The Regional Science Association Papers, 45: 67-85.

—— (1978) "Growth centers, rural development and national urban policy: a defense." International Regional Science Review, 3: 133-52

RODWIN, L. (1978) "Regional planning in less developed countries: a retrospective view of literature and experience." International Regional Science Review, 3: 113-131.

SALAU, A. (1980) "Evaluating the impact of administrative decentralization on regional development in Nigeria as an alternative growth center strategy." Ph.D. dissertation, Rutgers University.

SLATER, D. (1977) "Geography and underdevelopment." ANTIPODE 8.

———— (1975) Underdevelopment or Revolution. New York: Monthly Review Press.

STÖHR, W. (1979) "Evaluation of some arguments against government intervention to influence territorial population distribution." Presented at a United Nations Fund for Population Activities workshop on Population Distribution, Bangkok, September.

———— (1975) Regional Development Experience and Prospects in Latin America. The Hague: Mouton.

———— and H. PÄLME (1977) "Centre-periphery development alternatives and their applicability to rural areas in developing countries." Presented at the Latin American meeting, Houston.

STÖHR, W. and F. TÖDTLING (1978) "An evaluation of regional policies experiences in market and mixed economies." pp. 85-119 in N. Hansen (ed.) Human Settlement Systems. Cambridge, MA: Ballinger.

SUNKEL, O. (1969) "National development policy and external dependence in Latin America." Journal of Development Studies: 23-48.

United Nations (1980) Patterns of Urban and Rural Population Growth. New York: United Nations.

———— (1979) "Policies on human settlement in Latin America." Presented at the Latin American conference on Human Settlements, Mexico City, November.

———— (1977) Indicators of the Quality of Urban Development. New York: United Nations.

United Nations, Fund for Population Activities (1980) Rome Declaration. United Nations Fund for Population Activities conference on Population and the Urban Future, Rome, September.

UTRIA, E. D. (1972) "Some aspects of regional development in Latin America." International Social Development Review 4: 42-56.

VINING, D. R. and T. KONTULY (1978) "Population dispersal from major metropolitan regions: an international comparison." International Regional Science Review, 3: 49-73.

WEEKS, J. (1975) "Policies for expanding employment in the informal sector of developing countries." International Labor Review 3: 1-14.

WILLIAMSON, J. G. (1965) "Regional inequality and the process of national development: a description of the patterns." Economic Development and Cultural Change 12: 3-45.

World Bank (1980) World Development Report, 1980. Washington, DC: World Bank.

Part IV

The Local State:

Capital Accumulation and Political Conflict

8

Restructuring the American City:
A Comparative Perspective

NORMAN I. FAINSTEIN

SUSAN S. FAINSTEIN

☐ UNTIL MIDWAY IN THE last decade two metaphors captured the physical contrast between major European and American cities: a shallow bowl and a doughnut. European metropolises displayed a preserved historic center, serving expensive consumption, surrounded by high-rise commercial and residential development (Lichtenberger, 1976). The American city, in contrast, revealed a partially abandoned, desolate core, dominated by low-income, minority populations threatening an increasingly beleaguered central business district. Dynamic commercial, residential, and industrial development occurred on the outskirts, each enriching the other. The two modes of urban development had different impacts on social inequality (Fainstein and Fainstein, 1978b). In Europe, the greater physical intermingling of rich and poor within the same political boundaries minimized the use of local jurisdictions as bulwarks of privilege on the outskirts and dumping grounds for problems at the core. In the United States, reliance on local taxation and the concentration of impoverished minorities in inner cities exacerbated their already deprived status.

Suddenly, however, the U.S. picture appears to have changed, at least in a number of prominent places such as New York, Boston, San Francisco, and Denver. These cities have witnessed massive new public and private investment in the core, resulting in displacement of the poor and transformation of function. Downtown has become a source of profit once more as its uses have changed — the factory, the port, and the working class district have been replaced by the office, the tourist center, and the upper-class neighborhood. While physical appearance does not yet correspond to the European model, the domination of the city center by corporate and upper-class commercial and consumption uses appears to indicate a convergence in the social uses of space.

To the extent this convergence continues, students of urban development must necessarily revise their conclusions about the evolution of U.S. cities. Previously we saw a correlation between state policy on the one hand, and the spatial expression of social inequality on the other. In an earlier work (1978b), we attempted to show how weak U.S. urban policy and land-use controls were in part responsible for the characteristic physical and social form of the American city. We assumed that this pattern would continue. But we now think that the connection between the urban picture and social outcomes is becoming more complex and differentiated. It is important to explore both why this is so and what its implications might be for class and racial inequality.

Our contention here is not that most American cities now conform to the dominant European type,[1] but rather, that convergence, where it does happen, has important theoretical and political implications. We therefore (1) describe the nature of present convergence and the American locales in which it is occurring; (2) briefly summarize the proximate causes of former differences, then discuss the worldwide and uniquely American forces currently molding city form so as to explain present similarities; and (3) analyze the consequences of American changes for class and racial groups and the kinds of ensuing political conflicts precipitated by altered urban function. In our discussion, we use the term *conversion* to refer to changing urban class composition and economic function; *convergence* means increasing resemblance to the European model.

THE CHARACTER OF CONVERGENCE TOWARD THE EUROPEAN MODEL

Urban development gives physical expression to social inequality, both in the quality of the built environment and in the spatial distribution of the rich and the poor, of the socially dominant and socially marginal. Urbanism also represents a means of collective consumption, whether in housing, neighborhood life, or cultural commodities. The U.S. path of urban development continues, overall, to differ from the European in terms of each of these dimensions. But the aggregate trend masks the increasingly uneven character of change, one form of which constitutes a redirection of development toward the European model. Since this redirection is necessarily superimposed upon an already-formed built environment, change has been most rapid on the social side, in a transformation of the social uses of city space. We will identify the extent and nature of this convergent tendency by moving the discussion through three levels of analysis: (1) national aggregate, (2) alternative types of cities — "old" and "new", and (3) the general character of conversion within the "old" type. Because the first two levels of discussion dominate the literature, it is our intent to summarize some recent evidence, then to look more closely at old cities undergoing conversion.

AGGREGATE TRENDS AND ALTERNATIVE TYPES

The forest encompassing the various urban trees shows two dominant trends over the last decades. First, there has been a modest improvement in the absolute well-being of the lower classes and little deterioration in their *relative* economic position, both primarily the result of state income transfers and welfare expenditures (Plotnick and Skidmore, 1975; Plotnick and Smeeding, 1979; Danziger, Haveman, and Plotnick, 1981). Second, the social composition of central cities has become increasingly skewed toward lower-income groups, not so much because the poor live only in such places as because middle-class people have moved out either to metropolitan rings or to entirely new locales. In other words, the U.S. welfare state has expanded its activities almost to European levels — although expenditures are more erratic here (OECD, 1978) — and has established a posttransfer floor which has maintained the position of the lower classes in spite of increasing inequality in private market incomes. Social or vertical inequality has, therefore, taken on a European cast. Yet spatial inequality further reproduces the American picture of the huddled masses occupying the hole in the urban doughnut.

There are strong indications of worsening *urban* inequality *in the aggregate*. Between 1960 and 1978 central cities (CCs) lost 9 percent of their white populations, while their black populations grew about 40 percent (USDHUD, 1980: 1-14). By the end of the decade, 55 percent of blacks versus 24 percent of whites lived in central cities (USDHUD, 1980: 1-15), and almost one-third of the entire central city population was black or hispanic (USDHUD, 1980: 1-3). During a period (1969-1976) in which government-defined poverty dropped for the United States, it increased by 6 percent in CCs and by 16 percent in CCs larger than 1 million people (USDHUD, 1980: 4-2), with the proportion of CC blacks in poverty rising at similar rates — to the point where 31 percent of blacks living in CCs of any size as of 1976 were defined as poor (USDHUD, 1980: 4-4)

Central cities became impoverished both relative to their recent past and to their ring suburbs. Thus, Sternlieb and Hughes (1981: 54) estimate that the effect of outmigration of higher-income households and inmigration of lower-income households produced an aggregate loss of CC household income of $65 billion between 1970 and 1977 (using constant 1976 dollars); this translated into an average annual withdrawal of more than $9 billion in consumption capacity, with perhaps one-quarter to one-third of that decrease in buying power felt in the housing market alone. Not surprisingly, the gap between CC and suburban family income increased appreciably during the 70s (USDHUD, 1980: 4-17). These and many other indicators reflect the continuity — and divergence from Europe — of the aggregate tendency of U.S. urban inequality. In the words of two authorities:

> The urban crisis is not over — it is rather entering on its most fearful challenge. The demographic shifts within our society have left major urban areas increasingly the focal point for the distressed — not merely the impoverished, but the increasingly impoverished (Sternlieb and Hughes, 1981: 55).

Alternative types

But the aggregate picture also shows regional shifts of population toward the so-called sunbelt, where "post industrial" cities are growing in size and prosperity. In the overall context of what Berry (1980) calls "counterurbanization" and a still-born national commission termed the "deconcentration of urban America" (President's Commission, 1980), such cities as Houston, Dallas, Phoenix, and San Diego constitute a distinctly different type from the Clevelands and Detroits. These new cities have no clear urban core or at most an occasional island of highrises on a flat plane of moderate to low density. They are automobile and homeowner cities. They have modern, clean manufacturing (electronics, aerospace industries), a very large service sector, expanding tax bases, relatively high median incomes, and the absence of sharp urban/suburban inequality, as outlying areas are usually annexed into the jurisdiction of the central city.

The new cities, by contrast with the old, are financially solvent (Muller, 1981). The rise of the new cities has been based not only on their ability to capture expanding sectors of the postindustrial economy, but also on the political hegemony of their business classes. Relative fiscal solvency stems from low taxation, low expenditure regimes, and produces greater levels of social inequality than would be tolerated in the old cities of the North (see, inter alia, Perry and Watkins, 1977: 277-305; Mollenkopf, 1981). Thus, the new American cities differ sharply from the European model in their physical characters and in the low social wage they pay their lower classes. Given the preference of the Reagan administration for local determination and financing of social expenditures, we may expect further divergence in the coming years.

The general picture of old cities continues to show deterioration by every possible indicator, as might be supposed from the aggregate data presented earlier; for, when the bright new cities are subtracted out, the remaining old cities leave a bleak visage. But there is a countertendency, a *converting* old type of city, which is moving in the direction of the European model. Because this tendency usually involves extreme uneven development within cities and because it is very recent, it has not yet shown up definitively in aggregate data, even at the level of individual cities. There is, however, much suggestive evidence in the professional and popular literature that major changes are taking place in such old cities as New York, Boston, Chicago, Minneapolis, Pittsburgh, Denver, and San Francisco.[2]

THE CONVERTING TYPE

Almost every old city in America is trying to convert itself. Everywhere we find the familiar symptoms: the convention centers, government headquarters, downtown hotels, sports arenas, and office complexes through which officials and capitalists attempt to capture the dollars of the service economy. But in only some old cities has there been sufficient expansion of service activities combined with investment in real estate to produce a *positive* transformation in core land use. Instead of islands of hopeful development (most of which are publicly funded) surrounded by acres of parking lots, converting cities exhibit a new urban core with a complete connecting fabric of structures and activities. The difference between old cities in general, and the subset undergoing conversion, is created by the interaction among the three factors of demography, housing, and occupation *within the context of increasing private investment,* which facilitates major changes in the utilization of land in the urban core, as well as in other growth nodes around the city. Thus, *converting cities*

*are defined by transformation of core land use in a positive invest-
ment situation.*

It is our intent to describe the typical character of the conversion
process in such places as Boston, New York, and San Francisco,
where it is self-evident, rather than to measure the extent of conversion in these or other old cities. The following pages, then, fashion an
ideal type of process and structure, fleshed out with some evidence
about New York City.

Conversion is quick and difficult to predict. Underlying economic
forces reflecting changes in the American and world economy are
catalyzed by particular events in a disequilibriated process of development. Governments and the local bourgeoisie are often surprised by events. Thus, New York City officials tried to keep artists
from living in a vacating industrial district of Manhattan (SoHo), yet
those artists became the vanguard for a rapid infusion of commercial
and residential investment. San Francisco decision makers opposed
rapid transit stations in the heart of the Hispanic ghetto (Mission
District) because they wished to keep the lower classes off the new
"suburban-oriented" trains. They thought they had lost when they
capitulated to community political pressure and built two Mission
stations. But, ironically, the stations proved to be growth nodes for
the gentrification of the Mission and will, in years to come, service its
new middle- and upper-class residents.

Change is uneven. Alteration not only takes place in fits and
starts, but now here and now there. Unevenness is exhibited in the
way property rises and declines in value across the urban geography
(Harvey, 1974, 1977). Areas of redevelopment activity keep attracting
investment even as prices are bid up rapidly, while nearby locations
languish. Core areas gentrify while peripheral neighborhoods experience arson and abandonment, perhaps antecedent to new development. In a social sense, conversion is also uneven to the extent that
the new prosperity heralded by city officials and the upper classes is
not accompanied by improvement in the material condition of the
lower classes, who experience revitalization in higher transportation
costs and loss of such strategic advantages as derived previously from
concentration on prime real estate.

A new fabric is established for the built environment of the core.
Construction and rehabilitation is directed toward office buildings,
upper-income residences, and specialized consumption activities.
Isolated, "defensible," office towers are joined by luxury highrises
and a network of shops and restaurants. Manufacturing space in core
locations is converted to either offices or expensive housing. The

existing architecture of the core is, in effect, rehabilitated for service production or for consumption by the upper classes.

Core areas are converted to new consumption functions. Both as cause and effect of the changing demographics and market situation of converting cities, the character of core merchandising changes. The core is no longer the purveyor of mass market consumables, but rather of specialized goods and cultural activities. These commodities are aimed at the particular interests of fractions of the middle and upper class: travelers, tourists, and life-style groupings which require critical masses in order to permit individual consumption — e.g., homosexuals, singles seeking mates, and so forth. The manifestation of these new core consumption functions may be seen in rehabed old markets, dockside restaurants, small and highly specialized shops, hotels, resurrected "historical" districts, singles bars, restaurants with tin ceilings, plants, fans, and — as the food reviewer comments — well-meaning if unprofessional waiters in blue jeans.

The lower classes are displaced from strategic areas. During the fifties and sixties the first stages of this process were conducted largely under public auspices through urban renewal and highway schemes. In the present epoch, however, the "private" market forces enumerated so far combine to produce large-scale displacement. Lower-income and racial-minority groups are forced from the first ring surrounding the converting core as well as from gentrifying nodes in variously located residiential neighborhoods. Lower-class areas of converting old cities may experience further "decline" and devalorization as they absorb this population. In some cities, the lower classes are being forced outside of the political jurisdiction of the core altogether, to the point where officials in Denver, St. Paul, and San Francisco seem genuinely puzzled as to what has become of them. The process of gentrification of core sites and deterioration of locationally peripheral neighborhoods typically may skip a few "defended" areas in which white ethnic populations attempt to exclude both lower-class displaced households and gentrifiers. Eventually, however, as in Boston's North End and Charlestown, even committed working-class families succumb to the profitability of selling out.

Housing is consumed by the upper classes through ownership. Demand for suitable middle- and upper-class housing increases as new rental construction is minimal, and previously middle-class areas are occupied by displaced lower-class and minority households. The "bourgeois" rental apartment house in desirable neighborhoods begins to disappear as landlords are able to realize their capital through conversion to cooperative or condominium forms of occupant ownership.

New York is a case in point. A new social geography is rapidly crystallizing as the city exhibits the characteristics described above. Manhattan south of Harlem is being cleared of its previously large proletarian populations and much of its middle class as well. This new core is surrounded by the four other boroughs, of which two (Brooklyn and the Bronx), roughly speaking, house the working- to lower-class population, and the other two (Queens and Richmond) the working- to middle-class segment. A very few of the innermost sections of Brooklyn (Brooklyn Heights, Park Slope) have also been gentrified and constitute part of the core. The headquarters of service and financial corporations dominate the productive side of the core economy, while tourism, business travel, and culture consumption expand from their previously established bases. With its *grands ensembles* of public and publicly subsidized apartments situated primarily outside the Manhattan core, New York City takes on more and more the social geography of Paris, if not its central physical appearance.[3]

Evidence is accumulating about changes in demography, housing, and occupational structure which become expressed in core land use. The most important demographic phenomenon is decline in households with children and their replacement by those with adults in the labor market.[4] Changes in household size express themselves unevenly across the city, with the extraordinary outcome that by 1978 more than half (51 percent) of all renter households in Manhattan were occupied by a single person, an increase of 10 percent since the mid-1960s (Marcuse, 1979: 50). Even while the demand for housing remains acute, landlords abandon their buildings in declining neighborhoods. As of 1979, the city owned about 13,000 buildings acquired through tax default — most in Brooklyn and the Bronx — and faced the prospect of soon possessing 14,000 more, a total of roughly 30,000 occupied housing units. In contrast, the Manhattan core market is experiencing enormous levels of private investment. The seventies witnessed not only considerable new rental construction, but an escalating trend in building conversion. Perhaps 15,000 or more residential units have been created since 1970 by conversion from industrial uses. At the same time, cooperative schemes of ownership are rapidly replacing rental tenure for upper-income households, with anywhere from 20 to 30,000 units being withdrawn from rental status in the process (Roistacher and Tobier, 1980: 160-174).

The new patterns of residential consumption reflect the character of core production. The basic picture of conversion to service func-

TABLE 8.1   Changing Occupational Structure of New York City

EMPLOYMENT BY SECTORS (000s)	1969	(percentage)	1978	(percentage)
Manufacturing	826	22	492	15
Services (including banking and finance)	1247	33	1321	41
Government	547	14	540	16
Other	1178	31	931	28
TOTAL	3798	100	3284	100

SOURCE: The City of New York, Comprehensive Annual Report of the Comptroller for the Fiscal Year Ended June 30, 1980, p. 169.

tions is provided by Table 8.1, where we find a sharp absolute decline in manufacturing employment accompanied by both absolute and relative expansion of the service sector. An influential study (Conservation of Human Resources, 1977) identified a new productive agglomeration for the Manhattan core: the corporate headquarters complex. The vital sector of this complex is established by corporate service firms (law, accounting, banking, and so on) and corporate ancillary services (restaurants, hotels and the like). The complex accounts for 26 percent of private payroll in New York City for 1976. Much of the expansion of the service complex stems from banking, most noticeably foreign banking and investment operations, with assets increasing from $10 to $40 billion in just the first six years of the last decade. The shifting character of the New York economy is further evidenced by a decline from 25 percent to 14 percent of the manufacturing share in "export production" since 1960 (Drennan and Nanopoulos-Stergiou, 1979: 9). The loss of manufacturing employment in Manhattan results in the withdrawal of roughly 40 percent of building space from such uses (Roistacher and Tobier, 1980: 171; New York Times, March 8, 1981). Office buildings, in contrast, will expand by 25 million square feet during the first three years of the present decade — a private investment of more than $5 billion (New York Times, February 23, 1981). This development will produce a significant increase in Manhattan property values while the other boroughs remain stagnant or decline (New York City, 1981: 5).

Conversion, inequality and convergence. The New York experience is not unique. A similar pattern may be seen in Chicago. Smaller cities like Boston, Denver, and San Francisco are well along the path to wholesale conversion, to becoming the "Manhattans" of their metropolitan areas. How completely these tendencies will be realized, or how many other cities will move in this direction, remains unclear. But the character of inequality in the converting type is well defined. Large-scale capital accumulation in new service production facilities combines with gentrification to establish a revalorized and vibrant urban core (or core city). Working- and lower-class populations are forced to the periphery. Class inequality in no sense diminishes, yet its physical manifestations alter sharply.

Converting old American cities represent a convergence toward the European model in several ways. These include the new social functions of their cores, the elaboration of small-scale structures, gentrification and historical preservation, the elimination of private rental housing, and the redirected social gradient which places the upper classes back in the middle of town. In other critical respects, however, sharp differences persist — most obviously in the absence of much publicly provided housing for the lower classes and in the presence of acute intraclass racial conflict. Thus, convergence — like conversion — is uneven.

Both convergence and divergence between American and European situations are confined by the relatively constant trends of capitalist development and the sudden disjunctures propelled by reaction to crisis. To understand why American cities initially diverged from the European model in the post-World War II period, and why *some* but not most are changing course, we look very briefly at the determinants of the dominant U.S. trend, summarize contemporary worldwide factors producing similarities in advanced capitalist cities, and then identify the reasons why some American cities are reshaped by those worldwide (or "combined") forces.

CAUSAL FACTORS IN THE AMERICAN USE OF SPACE

Urban form and function throughout the advanced capitalist world respond to the law of combined and uneven development: There is simultaneous convergence and diversity as common economic forces are mediated in particular ways. The underlying capitalist contradictions between accumulation and consumption, as well as between organized production in the firm and disorganization in the larger society, manifest themselves as tensions and crises (see Beauregard, 1978). State policy and private decisions react to the unique events and choices that arise as economic relationships work themselves out in specific contexts. For example, Gough (1979) demonstrates that increasing levels of state welfare expenditures have been the common response of capitalist states to the consumption needs of the working class. The social welfare expenditure levels and policies of various countries converge. Nevertheless, different nations have adopted their policies at different times, depending on the balance of political forces within their territories, the pressure of proximate events such as war and migration, and a perception of crisis. The character of programs (e.g., fragmented versus centralized, targeted versus universal) reflects differing political institutions and administrative traditions. Similarly, convergence in modes of public and private urban investment and expenditure is subject to time lags, responds to ideological definitions of the meaning of events, and expresses itself in varying forms. Because cities constitute relatively stable, physical entities, patterns are less easily modified than is the case for other areas of investment and expenditure.

CAUSES OF THE DOMINANT TREND IN OLD U.S. CENTRAL CITIES

The reversals in urban fortunes described in the first section of this chapter were not predicted in the scholarly literature, although the aggregate trend was correctly forecast. Scholars characterized the old American city as a "sandbox" (Sternlieb, 1971) or "reservation" (Long, 1971), where poor people were isolated from the rest of society and given sufficient welfare benefits to defuse rebellion. They argued that the generative central city of the industrial era had reached its natural limits (Perry and Watkins, 1977: 277-305; Gordon, 1977), and consigned it to "pariah" status (Hill, 1978: 228-230; Castells, 1976) unless the obstacles to state planning and metropolitan tax base sharing could be overcome. Before we discuss the factors which countered previous trends in some places, we analyze the dominant mode so as to provide a context for our argument.

Principal among the causes of urban disinvestment was the increasing domination of central space by blacks and Hispanics, who, for a variety of reasons, could not be exploited at the same rate by industry and landlords as earlier generations of in-migrants. Because these populations were only a limited source of profitability for employers, merchants, and property owners, cities began to experience serious problems of both accumulation and realization. Labor elsewhere provided a higher return; tax revenues declined relatively and absolutely; retailing died; and the tenement landlord began to experience severe difficulty in meeting his costs. During the late fifties and early sixties, when minority groups were relatively quiescent, they were viewed by the governmental and corporate sponsors of urban renewal as drains on the economic base (Slayton, 1966).

Urban renewal programs, which sought to remove them and replace them with higher income residents, foundered, however, when private investors shunned the cleared land encircled by deteriorated neighborhoods — or worse yet, the empty lots surrounded by more empty lots (Kaplan, 1963). Whereas European governments at that time were constructing suburbs to house their working classes, the United States offered no similar program to coopt discontent over relocation. Continued efforts at clearing ever larger amounts of land were brought to a standstill during the middle sixties and early seventies, when urban riots and political movements (Fainstein and Fainstein, 1974) attacked the hegemony of the city officials and downtown business interests who comprised the "pro-growth coalition" (Mollenkopf, 1978: 141-145). Urban minorities thus limited the economic potential and threatened the political control of core locations; urban areas became the site of a legitimation crisis.

Simultaneously, changes in productive technology reduced the importance of urban agglomerations for industry and commerce (Sawers, 1975). Decaying public infrastructure and obsolete industrial plant contributed to the depletion of the core as economic decision makers refrained from the ongoing investment necessary for continued viability. Outmigration was further encouraged by antiurban cultural biases and federal tax and investment policies (Fainstein and Fainstein, 1978a; Markusen, Saxenian, and Weiss, 1981). The absence of regional planning controls (Sundquist, 1975), the relative cheapness of suburban and nonmetropolitan land, and the availability of lower-priced labor in the Southwest and abroad further encouraged the self-reinforcing process of flight by business and population.

CAUSES OF CONVERSION

The combined effects of population change and dispersal at first glance point to ultimate abandonment of the older city. But shifts in the structural context of urban development and a decline in racial militancy together enhanced the profitability of investment in the core under particular conditions. The consequence was a new cycle of investment and growth in those cities where a shift in economic function had occurred. The impetus to these new developments, which have produced in America the Europeanized metropolis, can be traced in part to the emergence in the United States of forces which are shaping land use patterns throughout the advanced capitalist world. *Receptivity to these forces depended on overcoming the factor –domination of the core by impoverished racial and ethnic minorities – that was the primary cause of problems of accumulation, realization, and legitimation.*

World factors. Common factors affect land use and economic concentration in all capitalist countries at similar stages of development. Foremost among these are the direct effects of the mode of production. The intermingling of ownership and control in the nineteenth century, combined with the growth of large factory work forces, created dense metropolises exhibiting extremes of wealth and poverty. In the present epoch, separation of ownership and control, production and administration, permits the rich to escape the poor, managers to avoid encountering their workers. The industrial city shows itself a highly transient phenomenon. Urban locations are resuming the classic economic functions of the preindustrial town — as administrative centers, marketplaces for producers and consumers, and hostels for travelers. These revived functions are manifested in new office buildings, a proliferation of stock and commodity exchanges, and vast commercial developments on the sites of old markets like Covent Garden, Les Halles, and Boston's Faneuil Hall Marketplace.

While capital and industrial labor no longer require close physical proximity, benefits of agglomeration still accrue to control and service functions. Production and circulation are freed from a particular urban form, and telecommunications can substitute for face-to-face encounters. Nevertheless, locations for meetings and the exchange of confidential information, as well as centers for legal, financial, and information services continue to be useful, though probably not essential, for capital accumulation. In the sphere of consumption, the wealthy owners of capital and the executive stratum need sufficient concentration to spawn the hotels, restaurants, and shops that cater to their buying preferences. While luxury tastes can be served within numerous geographies, their satisfaction in city cores, close to both executive offices and cosmopolitan hotels, is particularly convenient. In turn, upper-class consumption produces externalities that increasingly make the city attractive for middle-class uses, thereby enlarging the demand for urban space and raising the profitability of capital investments in real property (Lamarche, 1976).

The growth of tourism as one of the leading urban industries (Judd and Collins, 1979) combines in a novel way production and consumption patterns in the development of place. Place itself becomes an object of individual and collective consumption, causing producers (the tourist industry and the municipal government) to invest on a large scale in the production of aesthetically appealing areas (see Hartman, 1974). The expansion of air travel enhances the process by facilitating the placement of production sites in remote locations, thereby deconcentrating the working class and permitting its removal from the core. At the same time it brings wealthy consumers and business executives, including those whose headquarters have left the city, to urban areas. Usually the city center is a more attractive location for carrying on business and consumption activities than the service complex surrounding the airport. And, whereas the suburbs of major metropolitan areas may reflect the residential choices of these same travelers, they are not their preferred visiting places.

Governmental activity further determines the pattern of urban development. While American state intervention has certain unique aspects which will be discussed below, state participation is a significant force in shaping European and American cities. The state, in its mediating role, facilitates capital formation, planning, coordination, and social welfare (O'Connor, 1973). Its imprint on the cityscape is embodied in bridges and transit systems, zoning-induced uniformities and large-scale developments, parks and welfare centers. State intervention also sufficiently rationalizes competition in the urban property market to permit wholesale transformation of urban districts (Lojkine, 1976). The state's role is as direct investor in land acquisition and infrastructure, coordinator of sources of financing, relocator of former occupants, and subsidizer or builder of new structures. Even in those areas being upgraded through private investment in the existing stock, the state facilitates the process through regulation, tax abatement, and loan subsidies. But its fiscal limits in carrying out these functions catch it between tax revolt and service reduction, with a consequent undermining of the forces of development and stability (Hill, 1978; Alcaly and Mermelstein, 1976).

Common demographic, economic, and energy-related factors have also shaped the context in which advanced capitalist political economies are operating, with similar consequences for urbanism. Lowered birth rates and decreasing household size have increased the demand for urban residential units. Small — usually two-earner — households put pressure on the upper end of the housing market. With construction very expensive, however, only the wealthiest households can afford new housing. The remaining middle-class households rent or purchase units in rehabilitating neighborhoods previously occupied by the working-class. The considerable expansion of childless households in the urban demographic structure thereby has the double effect of providing families willing to live on the "frontier" of mixed-class neighborhoods and of maintaining demand for housing in such neighborhoods even while overall central city population and density drop.

New household formation in converting cities is rooted in an expanding service sector. The U.S. occupational structure has undergone two critical shifts in recent decades, both of which reflect trends common to advanced capitalist economies. First, the leading sectors of manufacturing are now to be found in high technology industries. Second, aggregate manufacturing expansion has slowed almost to a halt while the service sector has grown rapidly. The former trend is reflected primarily by manufacturing plants located in sub-urbs and new cities, as old cities usually do not have sufficient advantages for modern manufacturing production. Service occupa-tions, however, are expanding even in old cities. Thus, in Figure 8.1 we see a comparison of large central cities growing and declining in jobs during the period 1967-1972, samples roughly corresponding to new and old cities. We find that most of the loss (78 percent) in the old cities was accounted for by a declining manufacturing sector. In fact, these cities experienced absolute growth (13 percent) in service occu-pations. Cities able to take advantage of new service occupations, instead of merely being devastated by disappearing manufacturing, in effect redefine the functional mix of urbanism and thereby the charac-ter of the city under capitalism.

Several other forces affecting North America and Western Europe have pushed in the same direction. Escalating energy costs foster concentrated development, although improvements in com-munications technology exert a countervailing tendency (Pool, 1980). Worldwide inflation and expansion in the number of households make land in metropolitan areas more and more valuable; the use of real estate as a hedge against inflation heightens its cost relative to other commodities. Government restrictions on land use further escalate the price of land available for development. These elements all in-teract to make the most profitable piece of real estate a centrally located parcel developed to very high densities surrounded by other properties that do not threaten its prestige, physical security, and future sale.

The factors summarized here might establish an inevitable trend toward urban reinvestment. Yet, as we noted earlier, such cases are relatively unusual in the United State. Inner-city poverty and peripheral expansion continue to be the dominant American mode, as a consequence of the continued effects of minority occupation and capital disinvestment. We look now at the particular factors which have caused certain U.S. cities to respond to world trends and deviate from the American norm and also at the limits on these trends as they impinge on the American situation.

American particularism. Changes in American cities, far more than in Europe, result from a complex of uncoordinated initiatives.

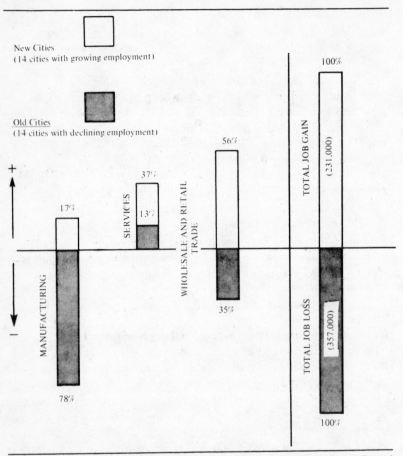

Figure 8.1 Composition of Employment Change for Twenty-Eight Large Central Cities, 1967-1972

The new downtowns are certainly products of government planning, but the thrust of that planning does not extend to limiting peripheral growth and competition, or providing sufficient new housing for displaced residents. As a consequence, the transformed cities, in which private investment greatly exceeds public activity, exist where the interaction of a number of variables, some fortuitous, have produced the opportunity for large speculative gains.

The pattern is now familiar. Once a bridgehead of upgraded territory is established, the devalorized property of the old central cities is sold at large multiples of its original buying price. Since private developers are not restricted by the stringent relocation requirements that gradually resulted from the urban renewal battles (Sanders,

1980), the process of transformation can proceed extremely rapidly. Former land uses, even those that were moderately profitable, become obsolete as the rate of return falls far below that achievable through turnover. High interest rates mean that any new investment must aim at a very large return. In the words of a Miami Beach hotel association executive who is resisting the establishment of an historic district that would limit high-density redevelopment, "You don't make money running a hotel. Money is only made buying and selling. You put this Art Deco scam on Miami Beach, and you're going to lose investors." (New York Times, Feb. 26, 1981: A14). Eventually, when the conversion process has run its course, and an area has restabilized, speculative gains and heavy financing costs become incorporated in new property values.

For this pattern to impose itself on a particular city, especially in more than a few isolated enclaves, a combination of circumstances must be present. In San Francisco the process of government-sponsored urban renewal frequently faltered when confronted with highly mobilized resistance. But an element in that resistance was middle-class preservationists, who blocked the demolition of Victorian homes and proved to be the vanguard of a vast movement in private rehabilitation. Eventually wholesale transformation arose from the confluence of public clearance of low-income residents from areas surrounding the central business district (CBD), enormous inflation in the entire California real estate market, private rehabilitation and realization of windfall profits from resale, and the rising importance of the Bay area as a financial center for trans-Pacific trade.

Manhattan's recent spectacular real estate inflation owes little to governmental activity. Rather, an influx of foreign money, the increased importance of its financial institutions during a period of economic instability and heightened activity in financial markets, and the continued growth of information and communications industries offset the consequences of the much-lamented flight of corporate headquarters and manufacturing. Manhattan's isolated geography buffers it from the City's low-income inhabitants; protection of the core does not require their displacement by the urban renewal authorities, who, along the European model, mainly direct their efforts at the outlying boroughs.

Denver's early efforts at urban renewal produced a checkerboard CBD of highrises and parking lots. Substantial single-family homes within a short commute to downtown offices sold for modest prices. Entertainment and retailing clustered along highway strips within and outside the city boundaries. Suddenly energy-related firms (exploration, drilling, financing, and legal, as well as the coal and oil companies) became the most profitable and fastest-growing industrial sectors in the United States. Denver had traditionally been the center for many of them. Expansion of these firms caused the new office buildings to fill up, more to be built, and an influx of professionals and executives to bid for conveniently located houses.

The foregoing examples show an escalation of redevelopment when exogenous forces create a potential demand for central space. But the prerequisite for the process is the absence, whether preexisting or forced, of low-income inhabitants. Obviously, this prerequisite already exists in unpopulated suburban and exurban areas, which, moreover, once had the advantage of relatively low-priced, accessible land not requiring the eviction of occupants. A shift in this latter factor of price advantage, however, began to favor reinvestment in some cities. On the one hand, continued disinvestment made central city land relatively inexpensive; black migration from the rural South was ceasing; and urban land was already served by streets, transit, and utilities. City officials were happy to welcome private developers, especially if they were willing to rehabilitate old structures. Instead of requiring that the developer provide parking spaces, sewers, and road access, they used public funds to subsidize infrastructure. On the other hand, while cities were actively facilitating the development process, suburbs were increasingly zoning out new construction, as earlier occupants began to fear that higher densities would depress property values, increase taxes, and diminish amenities. Suburbs began to demand that the developer assume more and more of the responsibility for mitigating the public impact of new construction. Even when suburban zoning boards were compliant, the process of achieving approval became so attenuated as to encourage developers, who were carrying increasingly onerous financing burdens, to explore other places. Ultimately suburban exclusionism in regions like California, Colorado, and the New York metropolitan area began to have effects similar to European restrictions on fringe development, even though the two systems of control differed strongly in impetus and content. Significantly, in much of the American Southwest, where zoning is less stringent or nonexistent, metropolitan growth continues on the earlier model of peripheral expansion.

POLITICS OF CONVERSION

Post hoc, explanations of specific urban trajectories are over-determined — we can always identify a set of causal factors which are more than sufficient to have produced observed outcomes. Prediction, however, is difficult. If change in urban form and the spatial expression of inequality is assumed to be a steady, even, and continuous process, prediction is frequently wrong. Yet explanations emphasizing unevenness and discontinuity appear idiosyncratic, the more so as political factors become central to the analysis of cause and effect. One way to take a middle ground is by employing a general framework or model which can be applied to the recent history of U.S. urban development. We find such a model in the concept of systemic crises in capitalist political economy.

Urban development takes sharp turns depending on the particular resolution of system crises. These, in turn, reflect the interaction of evolving objective conditions, mass consciousness, and political action. Habermas (1973) enumerates the major types of capitalist crisis. His typology can be reformulated for our purposes into the following four categories: (1) direct class conflict within the realm of production; (2) accumulation crisis, evidenced in lagging productivity, obsolescence of plant and infrastructure, governmental insolvency, and diversion of savings into unproductive speculation; (3) realization crisis, manifested in insufficient aggregate demand and declining profitability; and (4) legitimation crisis, reflecting withdrawal of voluntary public compliance with dominant norms. Within the urban realm examples of these types of crisis are respectively: (1) strikes by municipal employees; (2) fiscal crisis and service breakdowns; (3) unemployment, declining tax bases, and bankruptcy of retail establishments; and (4) urban political movements and riots. In addition, response to national economic crisis produces urban consequences, as in the construction of low-income housing, either to defuse general labor militance (direct class conflict), increase employment (realization crisis), or meet lower-class expectations (legitimation crisis).

NATIONAL LEVEL

American urban policy has been traditionally inconsistent. Whereas European policies of housing and urban development formed part of the "post-war settlement" between capital and labor, American policy has lurched between poles of heightened and reduced intervention. There has never been a long-term commitment to low-income housing, planning, or tax base sharing. Forms of governmental intervention veered from the urban renewal emphases of the fifties and early sixties to enlarged social welfare spending in the next period and a seeming concession of urban space to lower-class populations. By the mid-1970s, however, the shift in political balance caused by the national electoral triumph of conservative forces and a decline in racial militance permitted a redefinition of urban need in terms of fiscal crisis. The structural conflict between capitalists and urban residents ceased to be defined as a question of legitimation requiring the cooptation of dissidents and was again seen as a problem of accumulation and realization. The 1960s community action and Model Cities programs were displaced by "revitalization" efforts aimed at CBD development and stabilization of moderate income areas.

Moreover, state policy toward cities is not only a matter of specific programs for housing and community development. Present influences on urban development derive primarily from general state response to the overall economic and social situation; that response, which aims to heighten capital accumulation and reduce social consumption, is a product both of objective economic conditions and a wider crisis of legitimacy than was embodied in the inner-city conflicts of the 1960s (Castells, 1980). Most Americans view growth, not social welfare expenditures, as the principal justification for maintaining the capitalist system and its derived inequalities. The contemporary effort to reduce the size of the federal budget responds to the particular demands of capital for tax relief as well as the outcries of the working population, which attributes the decline in its real wages to government spending, welfare dependency, and seemingly consequent economic stagnation.

The effect of Reagan's policies on central cities depends partly on whether they do induce general economic growth, and whether or not their economic stimulus is at the expense of the service sector of the economy, which has been the principal component of urban economic conversion. On the production, or supply side, then, there remains the question of the composition of growth, and a further question of the geographical impacts of increased defense spending. The even more problematical issue is the nature of public consciousness if Reagan's policies do not promote growth or if their costs fall so heavily on low-income people as to generate again the disorder of the sixties.

Within the country as a whole, Reaganism represents a shift from redistributional to accumulation programs and a likely worsening of inequality. Within cities the Reagan withdrawal from direct aid to local economic development efforts, the new emphasis on industrial production, and possible increases in unemployment may halt the process of conversion to a service economy and convergence to the European model that form the topic of this chapter. On the other hand, it is also conceivable that the momentum of conversion is sufficient to continue the process under changed circumstances. Lacking the gift of prophecy, we devote the last section of this chapter to a discussion of the present politics of converting cities and its effect on the distribution within them.

LOCAL LEVEL

The particular definition of political conflict that prevails at any point does not preclude the simultaneous existence of other definitions nor the possibility of reversion to earlier modes. Those cities which are undergoing or have completed changes in their economic functions and class composition display a different politics from their earlier selves, but the issues which characterized the preceding period have not been wholly resolved or displaced. Some of the present political situation stems from national factors that affect all cities; other aspects reflect specific struggles arising from changes in land use and function. We contrast here two political models — the declining city and the converting city — with the caveat that contemporary declining cities (e.g., Detroit, St. Louis) share characteristics of both models; we are, in fact, comparing the declining city of 1965-1974 with the converting city of today. We further qualify our argument by agreeing that struggles of the earlier time persist, though less intensely, in the contemporary converting city, and, in the event of further decline in the national economy, may reassert themselves.

Figures 8.2 and 8.3 present two differing models of politics in the declining and converting city, described according to situation, mode, issues, and outcome. The *situation* of the declining city is characterized by a racial (legitimation) crisis, efforts by government to mitigate social disorder through provision of jobs and services, and physical redevelopment financed and directed by government. Conflict occurs in the streets, in sit-ins and takeovers of government offices, in battles over occupancy of land and buildings. The politics of disinvestment determines who will control the shrinking pie. Clients struggle with bureaucrats over the membership and processes

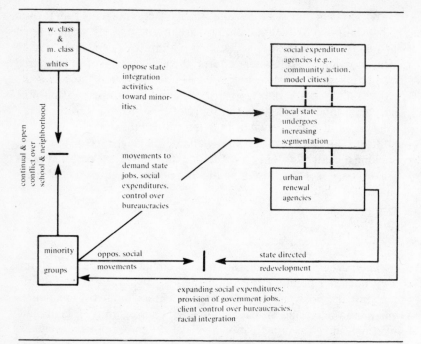

Figure 8.2 Dominant Form of Political Conflict, Declining Cities, 1965-1974

Political situation: legitimation crisis (racial definition); government efforts to integrate minority groups and simultaneously to displace them from core areas.

Mode: direct confrontation, politicization of socioeconomic conflicts.

Outcomes: increased social expenditures, integration of minority groups into local state, urban renewal of CBD and environs.

of governmental agencies that have client assistance as their ostensible aim. The *mode* of political influence is direct confrontation. The principal *issues* are the distribution and redistribution of government-controlled resources in the form of money, services, and jobs. Wins and losses are construed in racial terms. The *outcome* of struggle may be gains for low-income residents but in the context of virtual total dependency on the government sector and consequent vulnerability to externally generated changes in levels and categories of funding. Within the local political economy, opponents of governmental expansion may diagnose the outcome as a "cave-in" to pressure, resulting in a "negative business climate." This interpretation reinforces disinvestment in the private sector and heightens the significance of the government sector as "the only game in town." (This model is consistent with Castells' recent formulation [1980: 200-215].)

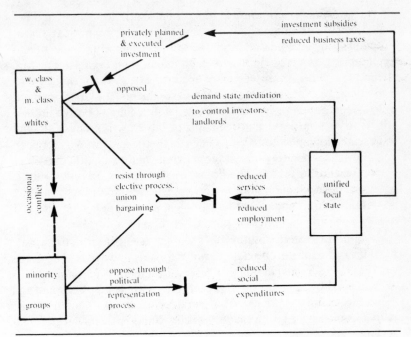

Figure 8.3 Dominant Form of Political Conflict, Converting Cities, 1975-

Political situation: accumulation/realization crises (fiscal definition), government efforts to improve fiscal management, encourage private investment.

Mode: privatization, shrinkage of political sphere; state mediates between capital and community; institutionalization of conflict in an (expanded) interest-group politics.

Outcomes: cutbacks in public employment, services, social expenditures; disciplining of municipal workers; displacement of w. & m. classes through private means.

The *situation* of the converting city develops from the ascendancy of fiscal crisis rather than legitimation as the preeminent urban problem.[5] The aims of government are directed at achieving improved fiscal management, dismissing redundant employees, and generally improving the investment climate, i.e., the conditions for accumulation. The *mode* of politics changes, as privately sponsored redevelopment replaces political allocation of resources. Politics reverts to the earlier, pluralistic form of interest group representation through council and bureaucracy. With the decline of mass mobilization, bargaining and routinized means of access once again predominate, although now with a larger number of "special interests" than in the long-ago days of consensual politics. Minorities seek representation through the electoral process and lessen their efforts to control bureaucracies directly through client representation.

As conversion continues, the problems of low- and middle-income residents change in nature, causing a shift in political *issues*. Previously, disinvestment produced a self-reinforcing logic of unemployment, poverty, housing abandonment, and tax base erosion. By contrast, private reinvestment establishes a cycle of rising property values, escalating rents and tax assessments, and displacement of low- and middle-income occupants. Politics now revolve around efforts to use the state as a regulator of the growth process. Racial issues are downplayed, as the lines of battle are drawn between neighborhoods and downtown, commercial and residential development, owner and renter. The controversies center on rent control, coop and condo conversion, historic preservation, and whether the emphasis of governmental redevelopment activities shall be directed toward downtown expansion or neighborhood stabilization.

The conservative solution to fiscal crisis, at both national and local levels, results in cutbacks of funds for social services. The needs of cities are redefined in physical terms. Community action programs are replaced by neighborhood preservation; funds are diverted from "soft" to "hard" services. Civil service unions dare not fight the cutbacks in services as taxpayer hostility limits their power to retain service jobs.[6]

The almost inevitable *outcome* is that the poor, and even the middle class, lose possession of the inner city. As in Europe, the bottom half of the population increasingly occupies peripheral areas (New York's outer boroughs; Daly City and Oakland, California; areas adjacent to highways strips in Denver), but different from Europe, receives few government services in recompense. The fortunes of particular displacees may be better or worse depending on whether they were bought out or forced out. But the locational choices of newly-formed low-income households are necessarily reduced, as there are simply fewer places where they can live. When, as in San Francisco and Boston, there is intense countermobilization of low-income people against the "pro-growth coalition," some benefits of growth accrue to them in the form of designated funds for housing; or, as in New York and Chicago, intensified efforts at community economic development.

The extent to which cities retain social services, protect the rights of occupancy of low-income populations, and provide alternative residential locals is a consequence of political struggle (Molotch, 1980). But the fact of conversion itself, while everywhere assisted by the state, cannot be produced by local political activity. Mollenkopf (1978) attributes urban redevelopment to the political triumph of the pro-growth coalition in particular cities. But every city has such a coalition, whose interests tend in the long run always to prevail (Stone, 1980). Such a coalition can create the preconditions for conversion, by clearing centrally located land and offering subsidies to developers. The actuality of conversion, however, has transpired in places where the coalition was subjected to intense countermobilization (e.g., San Francisco) and failed where it went unchallenged (New Haven).

The uneven quality of American capitalist development and the effects of deeply rooted economic forces are the principal causes of urban conversion in the United States. We have contended throughout this chapter that the result of the process is to make certain American cities resemble the European model in their social geography. But continued differences in the political and planning systems of the United States and Europe mean that partial convergence will have differing consequences for inequality. The structure of representation and the absence of centralized planning in the United States mean that questions of population migration and industrial location are never systematically addressed. The inadequate representation of low-income people, especially renters, at any level of government means that their interests are poorly defended.

The urban question, in converting cities, continues to be shaped by the racial and ideological fragmentation of the working class in confronting the forces of accumulation. As a consequence, low-income populations find themselves squeezed between the suburban ring and the expanding core in a period where public housing programs have dwindled almost to nothing. The position is particularly excruciating for blacks, whose race continues to exclude them from the suburbs. Conversion is thus a double triumph for capital, for it weakens political opposition through dispersal and it takes the problems of race and poverty out of the public view. Whether this victory is wholly at the expense of other groups depends on the concessions which may be won, and on the extent to which conversion also means expansion of employment and services — outcomes as yet unmeasured.

NOTES

1. We do not attempt to examine present European trends nor intra-European variations. Rather we assume a stable, ideal typical European model, perhaps best

captured by the examples of Milan and Amsterdam. There is considerable current debate concerning whether Europe is following the American trend toward decon-centration (Vining and Kontuly, 1978; Hall, 1980).

2. So far most of the discussion has been cast in terms of neighborhood change; see, for example, "The Revitalization of Inner-City Neighborhoods" (1980) and "Symposium on Neighborhood Revitalization" (1979). However, an excellent general review of evidence at a more macro-level is provided by Holcomb and Beauregard (forthcoming). See also the discussion of Boston in Menzies (1981).

3. Thus, while poverty increased in the city and in Manhattan as a whole between 1974 and 1977, so too did the median income of the borough. Although the other boroughs declined an average of about 15 percent in real income, Manhattan increased by 2 percent (Marcuse, 1979: 32, 55). The data indicate sharp uneven development at two levels, Manhattan versus the remainder of the city, and gentrifying Manhattan versus impoverished Manhattan.

4. For the city as a whole, the percentage of renter households (about three-quarters of all households) with more than two persons declined from 42 percent in 1965 to 33 percent in 1978 (Marcuse, 1979: 21). Correspondingly, one-person rental households grew from 27 percent of the rental total to 37 percent. Yet these changes did *not* result from increase in the elderly population, as the New York City population became slightly younger and the percentage of single-person households occupied by the elderly decreased from 35 percent in 1970 to 32 percent in 1978 (Marcuse, 1979: 47).

5. The definition of the New York City fiscal crisis clearly marked the hegemonic position of capital and its allies even while it was used as a political resource by conservative forces in many American cities (see Lichten, 1980). The politics of converting cities must be interpreted within the prevalent concerns of political elites, who almost everywhere now legitimize themselves with the active electorate by facilitating capital accumulation in the name of fiscal solvency. Thus, the typical politics of conversion manifests itself within a national definition of the urban problematic which both permits that politics (through mass *demobilization*) and itself arises from the objective conditions which make conversion necessary (the decline of industrial America).

6. In New York City the municipal unions are enmeshed in the ultimate structural discipline of being forced to invest their pension funds in city bonds, the value of which depend on the fiscal solvency of the city as defined by investment markets (see Lichten, 1980).

REFERENCES

ALCALY, R. E. and D. MERMELSTEIN [eds.] (1976) The Fiscal Crisis of American Cities. New York: Vintage.

BEAUREGARD, R. (1978) "Resolving tensions: planning theory about and for local planners," pp. 83-97 in H. A. Goldstein and Sara A. Rosenberry (eds.). The Structural Crisis of the 1970's and Beyond: The Need for a New Planning Theory. Blacksburg, VA: Division of Environmental and Urban Systems, Virginia Polytechnic Institute and State University.

BERRY, B.J.L. (1980) "Urbanization and counterurbanization in the United States." The Annals 451 (September): 13-20.

BURCHELL, R.W. and D. LISTOKIN [eds.] (1981) Cities under Stress. New Brunswick, NJ: Rutgers University, Center for Urban Policy Research.

CASTELLS, M. (1980) The Economic Crisis and American Society. Princeton, NJ: Princeton University Press.

——— (1976) "The wild city." Kapitalistate No. 4-5 (Summer): 2-30.

Conservation of Human Resources (1977) The Corporate Headquarters Complex in New York City. New York: Conservation of Human Resources Project, Columbia University.

DANZIGER, S., R. HAVEMAN, and R. PLOTNICK (1981) "How income transfer programs affect work, savings, and the income distribution: a critical review." Journal of Economic Literature 19 (September): 975-1028.

DRENNAN, M. and G. NANOPOULOS-STERGIOU (1979) "The local economy and local revenues," pp. 6-33 in R. Horton and C. Brecher [eds.] Setting Municipal Priorities 1980. Montclair, NJ: Allenheld, Osmun.

FAINSTEIN, N.I. and S.S. FAINSTEIN (1978a) "Federal policy and spatial inequality," pp. 205-28 in G. Sternlieb and J.W. Hughes (eds.) Revitalizing the Northeast. New Brunswick, NJ: Rutgers University, Center for Urban Policy Research.

——— (1978b) "National policy and urban development." Social Problems 26 (December): 125-146.

——— (1974) Urban Political Movements. Englewood Cliffs, NJ: Prentice-Hall.

GORDON, D. (1977) "Class struggle and the stages of American urban development," pp. 55-82 in D. Perry and A. Watkins (eds.). The Rise of the Sunbelt Cities. Beverly Hills: Sage.

GOUGH, I. (1979) The Political Economy of the Welfare State. London: Macmillan.

HABERMAS, J. (1973) Legitimation Crisis. Boston: Beacon Press.

HALL, P. (1980) "New trends in European urbanization." The Annals 451 (September): 45-51.

HARTMAN, C. (1974) Yerba Buena: Land Grab and Community Resistance in San Francisco. San Francisco: Glide Publications.

HARVEY, D. (1977) "Governmental policies, financial institutions and neighborhood change in United States cities," pp 123-140 in Michael Harloe (ed.) Captive Cities. London: John Wiley.

——— (1974) "Class-monopoly rent, finance, capital and the urban revolution." Regional Studies 8: 239-255.

HILL, R.C. (1978) "Fiscal collapse and political struggle in decaying central cities in the United States," pp. 213-240 in W.J. Tabb and L. Sawers, Marxism and the Metropolis. New York: Oxford University Press.

HOLCOMB, B. and R. BEAUREGARD (forthcoming) "Geographical aspects of urban revitalization." Association of American Geographers Resource Paper. Washington, D.C.

JUDD, D.R. and M. COLLINS (1979) "The case of tourism: political coalitions and redevelopment in the central cities," pp. 201-232 in G.A. Tobin (ed.) The Changing Structure of the City. Beverly Hills: Sage.

KAPLAN, H. (1963) Urban Renewal Politics: Slum Clearance in Newark. New York: Columbia University Press.

LAMARCHE, F. (1976) "Property development and the economic foundations of the urban question," pp. 85-118 in C. G. Pickvance, Urban Sociology. New York: St. Martin's.

LICHTEN, E. (1980) "The development of austerity: fiscal crisis of New York City," pp. 139-171 in G. W. Domhoff (ed.) Power Structure Research. Beverly Hills: Sage.

LICHTENBERGER, E. (1976) "The changing nature of European urbanization," pp. 81-107 in B. J. L. Berry (ed.) Urbanization and Counter-Urbanization. Beverly Hills: Sage.

LOJKINE, J. (1976) "Contribution to a Marxist theory of capitalist urbanization," pp. 119-146 in C. G. Pickvance (ed.) Urban Sociology. New York: St. Martin's.

LONG, N. (1971) "The city as reservation." Public Interest 25 (Fall): 22-38.

MARCUSE, P. (1979) "Rental housing in the city of New York, supply and condition 1975-1978." New York City Department of Housing Preservation and Development.

MARKUSEN, A. R., A. SAXENIAN, and M. A. WEISS (1980) "Who benefits from intergovernmental transfers?" pp. 617-691 in R. W. Burchell and D. Listokin (eds.) Cities under stress. New Brunswick, NJ: Rutgers University, Center for Urban Policy Research.

MENZIES, I. (1981) "Boston: qualified success." Planning 47: 14-22.

MOLLENKOPF, J. (1981) "Paths toward the post-industrial service city: the northeast and the southwest," pp. 77-117 in R. W. Burchell and D. Listokin (eds.) Cities Under Stress. New Brunswick, NJ: Rutgers University, Center for Urban Policy Research.

———— (1978) "The postwar politics of urban development," pp. 117-152 in W. K. Tabb and Larry Sawers (ed.) Marxism and the Metropolis. New York: Oxford University Press.

MOLOTCH, H. (1980) "The city as a growth machine: toward a political economy of place," pp. 129-50 in H. Hahn and C. Levine (eds.) Urban Politics: Past, Present, and Future. New York: Longman.

MULLER, T. (1981) "Changing expenditures and service demand patterns of stressed cities," pp. 226-300 in R. W. Burchell and D. Listokin (eds.) Cities Under Stress. New Brunswick, NJ: Rutgers University, Center for Urban Policy Research.

New York City, Office of Management and Budget (1981) The City of New York Financial Plan, Fiscal Years 1981-85.

O'CONNOR, J. (1973) The Fiscal Crisis of the State. New York: St. Martin's.

Organization for Economic Cooperation and Development [OECD] (1978) "Public expenditure trends." Studies in Resource Allocation 5 (June).

PERRY, D. and A. WATKINS [ed.] (1977) The Rise of the Sunbelt Cities. Beverly Hills: Sage.

PLOTNICK, R. D. and F. SKIDMORE (1975) Progress Against Poverty. New York: Academic Press.

PLOTNICK, R. D. and T. SMEEDING (1979) "Poverty and income transfers: past trends and future prospects." Public Policy 27 (Summer): 225-272.

POOL, I. S. (1980) "Communications technology and land use." The Annals 451 (September): 1-12.

President's Commission for a National Agenda for the Eighties (1980) Urban America in the Eighties. Washington, DC: Government Printing Office.

"The revitalization of inner-city neighborhoods" (1980) Special issue of Urban Affairs Quarterly 15 (June).

ROISTACHER, E. and M. TOBIER (1980) "Housing policy," pp. 145-80 in C. Brecher and R. Horton (eds.) Setting Municipal Priorities 1981. Montclair, NJ: Allenheld, Osmun.

SANDERS, H. T. (1980) "Urban renewal and the revitalized city: a reconsideration of recent history," pp. 103-126 in D. B. Rosenthal (ed.) Urban Revitalization. Beverly Hills: Sage.

SAWERS, L. (1975) "Urban form and the mode of production." Review of Radical Political Economics 7, 1: 52-68.

SLAYTON, W. L. (1966) "The operations and achievements of the urban renewal program," pp. 189-229 in J. Q. Wilson (ed.) Urban Renewal: The Record and the Controversy. Cambridge, MA: MIT Press.

STERNLIEB, G. (1971) "The city as a sandbox." Public Interest 25 (Fall): 14-21.

STERNLIEB, G. and J. HUGHES (1981) "New dimensions of the urban crisis," pp. 51-76 in R. W. Burchell and D. Listokin (eds.) Cities under stress. New Brunswick, NJ: Rutgers University, Center for Urban Policy Research.

————— (1975) "The declining and growing metropolis — a fiscal comparison," pp. 197-220 in G. Sternlieb and J. Hughes (eds.) Post-Industrial America: Metropolitan Decline and Inter-Regional Job Shifts. New Brunswick, NJ: Rutgers University, Center for Urban Policy Research.

STONE, C. (1980) "Systemic power in community decision making: a restatement of stratifiction theory." American Political Science Review 74 (December): 978-990.

SUNDQUIST, J. (1975) Dispersing Population: What America Can Learn from Europe. Washington, DC: Brookings Institution.

"Symposium on neighborhood revitalization" (1979) Journal of the American Planning Association 45 (October): 460-560.

U.S. Department of Housing and Urban Development [USDHUD] (1980) The President's National Urban Policy Report. Washington, DC.

VINING, D. R., Jr., and T. KONTULY (1978) "Population dispersal from major metropolitan regions: an international comparison." International Regional Science Review 3 (Fall): 49-70.

9

The Political Economy of Public Services

RICHARD C. RICH

"Those moments in which governing institutions appear as the direct, emphatic, and unmediated organs of a 'ruling class' are exceedingly rare, as well as transient. More often these institutions operate with a good deal of autonomy, and sometimes with distinct interests of their own, within a general context of class power which prescribes the limits beyond which this autonomy cannot be safely stretched" [Thompson, 1978].

The main job of municipal government is to create a climate in which private business can expand in the city to provide jobs and profit [New York Mayor Edward Koch, quoted in the New York Times, March 4, 1978].

□ URBAN DEVELOPMENT DEPENDS crucially on capital accumulation. Accumulation must precede all projects which transform urban space and its uses. Periods of urbanization in both Western Europe and in the United States have corresponded to phases in the process of capital accumulation which produce distinctive patterns of urban development (Castells, 1977; Gordon, 1976; Harvey, 1981). In capitalist systems the vast majority of the capital which supports urban development is accumulated in the private sector. Government action, however, underpins the commitment of capital to urban development in essential ways. Not only do governments accumulate and invest huge sums in the infrastructure of urban development (roadways, bridges, ports, water systems, and so on), but the produc-

AUTHOR'S NOTE: This chapter benefited from the critical comments of Norman and Susan Fainstein and Stephen White, and I am grateful for their assistance.

tion of "soft" services insures the usability of capital investment.
Hotels, apartments, and businesses, for example, would be of little
value without transportation nets to link them or police and fire
protection to provide minimal levels of security. One authority sums
up the relationship between public services and urban economic
development as follows: "Spending on schools, health care, higher
education and the like has come under increasing attack because they
are thought to be 'wasteful.' Yet, upon examination, it appears that
[these services] . . . have provided the basic economic infrastructure
upon which the post-industrial service city has been constructed"
(Mollenkopf, 1981: 108).

There are subtle and complex relationships between the public
production of services and patterns of urbanization in capitalist nations which are often neglected by students of urban service delivery,
but which shape that delivery in profound ways. In this chapter I will
attempt to outline the conceptual relations among public service
production, capitalist economics, and urban development patterns.
In doing so, I hope to lay a foundation for an enhanced understanding
of the place of locally provided public services in the broader context
of city politics and urban development — an understanding which
recognizes that patterns in the production and delivery of public
services are not only reflections of dominant economic and class
relations in a society, but significant shapers of those relations in
individual cities.

The argument has four parts. In the next section I trace the broad
outline of a theory of local service production by applying the concepts of critical or structural theories of the state to the activities of
local government. In the second and third sections I explore the
dynamics by which public service decisions are made and implemented in an effort to specify their relationship to larger political-
economic processes. The final section presents an analysis of the
interaction of service delivery patterns and urban development.

Without offering a purely functionalist interpretation, I do seek to
identify the functional relationship that exists among economic organization, public service production and delivery patterns, and
trends in urban land use. The thread which weaves these three concerns together is the relative political influence of the key actors
whom I will identify below — political influence granted *at least* as
much by structural position as by political organization and activity.
This relative power balance is reflected in the relations of control,
production, and distribution of services. I will explore each set of
relations in turn.

My argument is simply that public services bear a much more complex relationship to political power and development patterns than is commonly recognized by conventional, behaviorally oriented analyses. Recognition of the full character of the interface between service production and larger political processes leads to a set of expectations about service patterns and interpretations of observed patterns very different from those offered by behavioral analysis. We cannot adequately understand, and therefore cannot control, public service production until such time as we understand its place in the larger political economy. I hope, by this chapter, to encourage exploration of that place.

SERVICE PRODUCTION AND THE LOCAL STATE

Behavioral political science generally depicts public services as the product of a pluralist bargaining process in which various interest groups in the city vie for shares of the local budget and for control of its size and financing. (See Yates [1979] for an exemplary discussion, and Rich [1980] for a review of contemporary approaches to urban politics.) As a result, the provision of services which benefit relatively powerless groups are viewed as something of an aberration, and distributional patterns present a puzzle where population subgroups do not receive services in proportion to their political influence. (See Rich [1979] for a review of much of the literature on service distribution, and Rich [1981] for some more recent research on the subject.)

The most prominent traditions in the study of public services can provide only ad hoc and relatively unsatisfying explanations of observed patterns because they are not clear as to the purpose of public production of services. (See Peterson [1981] for a non-Marxist critique of behaviorally oriented studies of urban politics.) If the reason we have public services is addressed at all, it is generally presumed that they exist in order to solve human problems or create desirable conditions (ACIR, 1973; Lineberry, 1977). Such an assumption, however, leaves us without a basis for explaining many commonly observed facts about service production and delivery, except by ad hoc reference to such factors as inefficiencies in the organization of the public sector, imperfections in the demand articulation and interest representation systems in given cities, or the exceptional abilities of individual political leaders. Such factors are evoked to justify the provision of public services which are inadequate in both quality and quantity to solve the problem or create the condition at which they are aimed; to account for the provision of different types of services through systematically different financing arrangements — in short, to interpret all the deviations from what would be expected if services were produced in response to the decisions of a pluralist bargaining process and in order to serve societal interests (Mollen-kopf, 1979).

What is missing from this conceptual framework is any explicit theory of the institution through which public services are provided, the local state. Without a theory of the local state we have little basis for predicting patterns in city politics or urban development, because such outcomes are depicted as a product of the unique balance of power created by the skills and resources of the actors involved in given cities at particular times.

There has been much recent work on the theory of the state from a critical, frequently Marxist, perspective. This literature offers a point of departure in developing a theory of the local state which can further our understanding of public service production. Simply put, the state is the mechanism by which social and economic relationships are ordered through lawmaking and program implementation. At the local level, county governments, municipal governments, special service districts, and a variety of "special authorities" constitute the state.

While there is no single theory of the state to be found in the work of contemporary Marxists,[1] James O'Connor's (1973) identification of the functions of the state has been very influential, and it provides a useful basis for the present analysis. He depicts two (often contradictory) roles for the state in capitalist systems. First, the state creates the conditions for, and underwrites, private capital accumulation through "social capital" outlays of two types. These are (1) "social investment" projects that increase or insure labor productivity (e.g., utilities, transportation, job training), and (2) "social consumption" projects that lower the costs of class reproduction (e.g., education, social insurance, health care). Second, the state helps maintain social control by shouldering the burden of "social expenses" through projects which mitigate the social disorder created by the operation of the market (e.g., police protection, welfare transfers, nutrition programs, drug abuse control).

This conceptualization of state activity, when applied to local government, places service production in a larger context, and renders understandable observed patterns in the relations of control, production, and distribution of services.

THE CONTROL OF PUBLIC SERVICES

Virtually everyone agrees that public services are necessary. The debate surrounding their production centers on determining the appropriate mix and level of services (the size and composition of the public sector) and on selecting funding arrangements. In the choice of funding arrangements, U.S. and Western European cities operate in very different contexts. In the United States the most progressive and productive revenue sources are monopolized by the federal government, and most local governments are constitutionally prohibited from deficit financing of operating expenditures. European cities, by contrast, share more fully in the revenue base of their national governments (Bennett, 1980). The European pattern produces more consistent *national* urban policies and greater fiscal capacity for cities, but local governments, themselves, exercise less control over the size and content of the local public sector than in the United States (Ashford, 1980).

In both systems, harmonization of the interests of the state and capitalist institutions is fostered by a financing system which removes decisions about social capital investments as much as possible from the political influence and attention of the working class, and serves to hold social expense outlays to the minimum necessary to insure the maintenance of order. Thus, in the United States, most locally sponsored capital improvements are funded by bonded indebtedness rather than direct taxation, and many are administered through politically invisible special service districts. By contrast, operating expenses like police and fire protection, sanitation, and social services are funded by highly visible local taxes, and become objects of intense political conflict among elements of the working class — conflicts which often pit racial groups against one another, and public employees against local taxpayers (Judd, 1979).

In either system, it is in the interest of capitalists to support a range of social investment by local governments. Indeed, both Europe and the United States have seen the mobilization of what Mollenkopf (1978) terms pro-growth coalitions of public officials and corporate leaders who pressed for significant public commitments to capital improvements and other social investment projects (Romanos, 1979; Stone, 1980). This investment socializes many of the costs of labor reproduction and infrastructure provision that would otherwise have to be borne by individual firms. As state production grows, however, it comes to play an increasingly important role in people's daily lives. Issues of effective and equitable service delivery become central to urban politics (Fainstein and Fainstein, 1974), and municipal employees unions gain increasing influence over budgetary decisions (Spero and Capozzola, 1973; Zack, 1973). The short-run interests of the public officials who manage the local state come to run counter to the interests of capitalists, and local budgets appear to be out of control. State investment may siphon off so much capital from production, circulation, and accumulation that recession and stagnation threaten, and local governments are faced with revenue shortfalls, causing fiscal crises. Capitalist interests, especially finance capital, seek to avoid such events by continuing pressure for limited services, and they press for service cutbacks once fiscal distress develops (Morris, 1980).

Local governments on either continent have little choice but to comply. In the politically fragmented U.S. system, the mobility of firms grants them a capacity to move away from those jurisdictions that tax too heavily or provide a mix of services which businesses find disadvantageous. By threatening to move and playing one jurisdiction off against another, firms can essentially dictate the size, scope, and financing of the local public sector in the long run (Newton, 1976; Peterson, 1981). Where hard-pressed city governments attempt to shift operating costs into the capital budget to avoid service cuts, the control which finance capital exercises over the bond market can result in a loss of credit rating that further exacerbates the fiscal crisis (Twentieth Century Fund, 1974; Sbragia, 1981).

Even in the more politically centralized countries of Europe, firms enjoy this power of fiscal blackmail. Tax policies may be more uniform across cities, and localities' service delivery obligations may be set at the national level, but firms still exercise great power by virtue of the link between their productivity and state revenues. Not only can firms move across national boundries, but politicians know that to let them wither in place by taxing away their competitive edge or failing to provide adequate infrastructural support would rebound to the detriment of state revenues. They may thus restrict public services in order to forestall the need to cut them even more severely at a later point (Gough, 1979).

At the most general level, it is capitalists who control city budgets, not local politicians, voters, or municipal employee unions. In pursuit of the economic interests of their individual firms, capitalists make choices which set limits to the size of the local public sector and dictate the mix of services it will contain. They need not conspire or intend to do this, or even be aware of the consequences of their actions. The structural domination of capital is a product of the interdependence that exists between the fiscal condition of local governments and the economic health of their home industries (Watkins, 1980).

The extent to which city budgets are held hostage to the profitability of local firms may be an artifact of the financing arrangements adopted for public services in a given nation (Sbragia, 1981). It is quite possible that European cities offer a much wider range and more fully developed set of public services than U.S. cities largely because European financing arrangements free the size and composition of each city's budget from the profitability of local firms by allowing them to tap into a national pool of revenue for major portions of their expenditures. In contrast, the extreme political fragmentation of U.S. urban areas offers opportunities for firms to gain advantages through geographic mobility as a consequence of heavy reliance on local financing of public services (Fainstein and Fainstein, 1978). These variations in fact, underscroe Theda Skocpol's (1980) persuasive argument that political forms exert an independent influence on the resolution of conflicts born of the capitalist mode of production. Nonetheless, relatively centralized financing of services does not alter the basic locus of control. Centralization may shift the focus of lobbying efforts and reduce regional variation in fiscal conditions, but it does not eliminate the link between private sector prosperity and public revenues in capitalist systems, and therefore leaves service delivery budgets essentially dependent on decisions made in the private sector.

This point is dramatically illustrated by the recent rise to political influence of advocates of "supply-side economics" in U.S. politics. (See Gilder [1981] for a statement of supply-side economics.) A major portion of their successful call for cuts in federal domestic spending and investment in "reindustrialization" is based on the logic of the Laffer curve (Laffer and Seymour, 1979). A basic component of this concept is the idea that beyond some level of taxation, tax increases so depress investment and productivity as to produce *decreases* in government revenue. This situation is quite compatible with the logic of state fiscal crises laid out by contemporary Marxists as cited above (see O'Connor, 1981).

The supply siders' budget cutting efforts have been born of a conviction that *national* taxation levels are harmful to economic growth and the competitive position of U.S. firms in the world market. The cutbacks, however, will be felt most keenly in cities, with a predictably negative impact on the size and scope of the local public

sector (Reischauer, 1981). The analogy between the United States in
this situation and more politically centralized European nations
should be clear. While centralization may give individual cities the
capacity to provide services that their own economic condition would
not support, it does not remove the link between capitalist accumulation and public service expenditures, and therefore does not alter the
ultimate locus of control of local public service levels. Had U.S.
urban policy been more centralized and urban expenditures more
centrally financed, the impact of the federal budget cuts on city
services would have been even greater.

Thus, the context of urban service delivery in Western Europe and
the United States is different to the extent that (1) the level of political
centralization of urban policy-making shapes service decisions, and
(2) greater working-class mobilization in Europe presents the state
with more acute problems of legitimation. However, cities on both
continents are constrained in service decisions by the inherent link
between the size and composition of the public sector and the process
of capital accumulation (Peirce and Hagstrom, 1981).

THE PRODUCTION AND DISTRIBUTION OF SERVICES

The relationship between state revenues and private economic
activity sets general limits on the size of local budgets, and dictates
budgetary balances which favor capital investment and its attendant
debt service costs over collective consumption expenditures
(Szelenyi, 1981; Research Planning Group, 1978). Within this broad
context, however, the pattern of services and the distribution of
services can vary considerably. They are shaped by the relations of
production and distribution of services.

The nature of public services, especially the labor-intensive ones
that touch people's lives most directly, is such that service delivery
personnel can exercise considerable power over both the quality of
services and cost of service delivery to the city (Jones, 1980). While
city council decisions that commit huge quantities of tax dollars to
debt service and maintenance of capital investments often go largely
unnoticed by citizens, the daily decisions of public agencies and
personnel that add incrementally to the cost of government can become the focus of intense political attention. Indeed, conflict between
unionized city workers and both outraged taxpayers demanding "efficiency" in service delivery and mobilized community groups demanding access to the public service employment that represents an ever
larger portion of the dwindling jobs available to minorities has be-

come a centerpiece of central city politics. (Brown and Erie, 1981; Hill, 1978).

In cities experiencing fiscal pressure due to rising demands for services and declining revenues, public employee unions are often depicted as villains, responsible both for increasing costs of services and inequitable distributions. Popular experience with public services is such that this ploy on the part of political officials is usually successful, leading two observers to conclude that "city politics under conditions of decline has relegated minorities, public employees, and other [service demanding groups] to inferior positions and has put political power in the hands of elites whose interests are often opposed to these groups" (David and Kantor, 1979: 215).

Public employees have few defenses against this characterization since (1) personnel costs constitute the largest single item in operating budgets, and (2) productivity increases that might offset wage increases are virtually inpossible in labor-intensive fields (Poole, 1980). Public service unions' demands for wage and benefit increases openly pit the economic interests of their members against the economic interests of taxpayers, leaving them in an especially weak position in systems where class relations have not been institutionalized in the public sector through full-blown collective bargaining arrangements permitting strikes and binding arbitration (Weitzman, 1979).

But how much do service bureaucracies shape service patterns? How much of the "rap" for declining quality and rising costs of services must they bear? Recent behaviorally oriented research on urban service distributions suggests that local bureaucracies play a crucial role in determining service levels and distributions. Beginning from a variety of perspectives, this research has tended to converge on "bureaucratic decision rules" as the most powerful explanation of observed service delivery patterns (Sanger, 1981). Bureaucratic decision rules are the formal and informal guidelines that direct service personnel in making the myriad choices they confront about service allocations both on a daily basis and more episodically. Such rules are generally justified in terms of professional standards aimed at insuring efficiency and effectiveness in service delivery, even though they may actually be adopted because they serve the interest of individual agencies by making their job easier or deflecting public criticism by providing a rationale for decisions.

While there is good reason to question whether service agencies are unified enough to make decision rules effective guides to action (Nivola, 1979), the impact of such rules would be to foster a distribution of services in accord with "politically neutral" criteria that are blind to recipients' class standing. Some students of urban services point to evidence that distributional patterns seem to favor different class and racial groups in different cities and service areas in an almost random pattern as support for the thesis that class conflict plays no important role in public service delivery, due largely to the power of bureaucracies to control the day-to-day distribution of services (Lineberry, 1977).

Such an interpretation is possible, however, only if one disregards the role of public services in the larger political economy. From the perspective of structural theory, evidence that some services are distributed relatively equally among population groups, or even in a compensatory fashion (going disproportionately to lower-income people) can be seen as evidence that class interests *are* being served by the service delivery system.

To understand this, we must return to O'Connor's (1973) conceptualization of the functions of the state in capitalist nations. He distinguishes among those programs which represent social investment, social consumption, and social expense. Social investment services (capital facilities and their maintenance) make a direct contribution to profit and accumulation and are generally considered to be "essential" or "basic" services. To be effective they must be available to everyone (though not necessarily equally used by, or beneficial to, everyone). Research on those services has generally revealed a tendency toward equal distribution (Boots, 1972; Levy, Meltsner, and Wildavsky, 1974; Antunes and Plumlee, 1977). Social consumption services (e.g., libraries, education, housing, health, welfare) support accumulation indirectly by insuring the reproduction of labor. To be effective in this, however, they must be targeted to specific groups, and indeed, research generally suggests such targeting (Gold, 1974; Lineberry, 1977). Similarly, social expense activities, which are designed to preserve the social control essential to capital accumulation, must be targeted to appropriate populations to be effective.

Behaviorally oriented scholars often come to the study of service distributions with the assumption that public services *should* be equally distributed, or at least distributed in proportion to some criteria of need. When they find deviations from such a distribution, they treat it as an aberration (Rich, 1979). Structuralist theory, however, presents a picture of the functions of public services that leads us to expect unequal distribution of services as the normal functioning of the state in capitalism.

To begin with, it is important to recognize that there are at least two broad dimensions to the concept of service distribution. The first is the allocation of public funds among alternative types of services. This decision, made in the budgetary process, determines which among an array of potential services the state will produce, and the level at which each will be provided. The second dimension of service distribution is reflected in the allocation of those services that *are* to be provided among population groups. This allocation determines who will get how much of available public services. While it is influenced by decisions in the budgetary process, this allocation occurs primarily in the administrative process, with all levels of the bureaucracy playing a role.

Of the two distributive decisions, the former has far more impact. Decisions about what services to fund (let me call them legislative decisions for convenience) set the broad framework within which all other distributive decisions are made, and set constraints on the impact of those decisions on people's lives. To illustrate, the decision about which of several eligible individuals is to receive some welfare benefit is of less social significance than the decision about what group shall be eligible for benefits; or, the decision about how to distribute bus routes about the city has less impact on the transportation mix in that city than the decisions that determine how many bus routes there will be and what carrying capacity they will have. In each of these examples, the more crucial decisions are those made at the legislative rather than the administrative stage.

In many instances the effective decision about who will benefit from public services is made before public resources are ever mobilized for service delivery. While "basic" public services (e.g., police and fire protection, sewers, street lighting) benefit a wide spectrum of the population, many services are so specialized as to provide direct benefits to only specific subgroups within the population. (Consider the clientele for senior citizens' centers, job training programs, day care facilities, public housing, or summer jobs for unemployed youths as examples.) Thus, when public officials select the mix of public services that are to be provided, they also, to a large extent, determine the distribution of benefits from public action. The implication of this is that the most significant politics of urban service distribution probably takes place in the local budgetary processes and (to the extent that localities respond to initiatives from higher level governments) in state and national budget allocations.

By choosing to concentrate public resources on those services that confer roughly equal benefits on a broad segment of the public, political elites can insure that the impact of public expenditure will essentially maintain or reinforce the existing distribution of wealth and opportunity. An examination of the distribution of city expenditures among functional areas in the United States (Bureau of the Census, 1979) shows clearly that officials have, in fact, selected this option by concentrating their funds on distributive rather than redistributive services, and on services for which there are few private sector substitutes. Such decisions restrict both the impact that public services will have on societal problems and the redistributive effect of public provision of goods and services even before any services ever reach the street.

If, as an increasing number of students of the subject are arguing, bureaucratic decision rules are the principal determinant of the "street level" distribution of locally provided services, the simplified critical theory of the state cited at the outset of this chapter is again useful in helping us see the political significance of public services. When political elites set the rules of the service delivery game through decisions made at the legislative stage of functional budget allocations, they have every reason to be content to leave the calling of individual plays to bureaucratic umpires. Indeed, it is to their great advantage to have routine distributional decisions made by professional bureaucrats guided by technically determined rules. This tends to depoliticize local state action and, in a society valuing professionalism and scientific management, to legitimate the social relations sustained by the resulting service delivery patterns, allowing the local state to fulfill important ideological functions in the capitalist system (Gold, Lo, and Wright, 1975).

The "proper" distribution of services is a crucial prerequisite to state action effectively performing its economic functions of supporting production and reproduction. While specific service decisions (e.g., the repair of a pothole or the handling of a complaint) may be subject to random forces, the broad distribution of public resources among service types and of services among population groups must *not* be random if it is to be functional for the economic system. Thus, we should not be at all surprised to find that low-income people are disproportionately the recipients of locally provided welfare or social services, or that poor neighborhoods have more police patrol, or public housing facilities, or low-interest housing rehabilitation loans. If the function of state action is to counterbalance the destabilizing human problems created by the operation of the market (Harvey, 1978; Hill, 1978), then it is logical for those populations among whom these problems are concentrated to be the targets of ameliorative state action. Again, empirical research has generally revealed such a concentration (Wolch, 1981).

State action not only *facilitates* class reproduction by maintaining social control, but also more directly contributes to that reproduction. Thus, we should expect to find, for example, that the poorest schools are concentrated in low-income and working-class neighborhoods while better schools dominate middle- and upper-income areas; that children of lower- and working-class families are "tracked" into blue collar-oriented curricula while middle- and upper-class children are designated as "college material"; that welfare institutions seem only to contribute to recipients' inability to escape poverty; that many services which would contribute to the economic mobility of the lower classes (adequate day care, effective job training, certain types of small business assistance) are not provided or not provided in adequate quantity or quality by local governments; or that middle- and upper-class persons are allowed to isolate themselves into political jurisdictions where they can escape heavy tax burdens that inhibit the accumulation of capital. Nor should we be surprised when we find that the unintended consequences of state actions seem to perpetuate the problems they are nominally intended to remedy, as when the siting of subsidized housing reinforces the concentration of low-income groups into neighborhoods where children acquire educational and economic handicaps, or public housing programs lead to the displacement and further impoverishment of the poor. All of these and many more phenomena are examples of the class reproductive functions of state action at the local level and illustrate the intimate link between the political and the economic systems (Bowles, 1975; Gorz, 1977).

URBAN FORM AND PUBLIC SERVICE DISTRIBUTIONS

The administrative aspect of service distribution (as distinct from the legislative aspect) determines which groups and individuals receive existing services. This allocation is made principally through decisions about the *geographic* distribution of service resources. The geographic spread of service benefits is especially politically significant in the United States because the extensive class and racial segregation that characterizes its cities causes decisions about the physical location of services to be generally equivalent to decisions about the class and racial composition of service beneficiaries.

Most U.S. research on service distribution has focused on the administrative face of the phenomenon and has examined it *within* given local jurisdictions. This research has produced an empirical picture of public service distribution in the United States as dominated by unpatterned inequalities. It has, however, largely ignored what is arguably the most politically and economically important aspect of urban service distributions, the distribution of services *among* jurisdictions within metropolitan areas. The combination of political fragmentation and class and racial clustering in U.S. metropolitan areas creates situations in which local jurisdictions (municipalities, counties, and, perhaps most important in this context, special districts) can "specialize" in service packages that cater to narrow bands of the class spectrum (Cox and Nartowicz, 1980; Newton, 1976).

Through judicious use of tools such as municipal incorporation, special districts, and zoning laws, middle- and upper-class communities can effectively isolate themselves from working- and lower-class persons (Danielson, 1976), thereby freeing themselves of much of the social cost of insuring order and providing for class reproduction. Their efforts in this regard produce a legal separation of public resources from public needs which, in the absence of *extensive* state and federal funding of public services, effectively limits the redistributive impact of state action and insures poor services for impoverished communities (Hill, 1974; Michelson, 1975).

While political fragmentation and the attendant separation of public resources from public needs may bestow short term benefits on some middle- and upper-class suburbanites and certain elements of the capitalist community (Cox and Nartowicz, 1980), it contributes directly to the fiscal crisis confronting so many jurisdictions (Research Planning Group, 1978). Fiscally stressed cities find it increasingly difficult to perform either of the central functions of the local state, and public services become both the focus of efforts to manage the political economy and the source of class confrontation as capital attempts to restrict the state's interference with the accumulation process by cutting back governmental activities.

The politics that ensue (the politics we observe in the United States and England today) are instructive. For capital, the secret to success in this case is in reducing the costs of the state's social control and reproductive activities to a minimum while preserving as much of the state's accumulation and production-supporting function as possible. The organizational structure of local government in the United States is well-suited to this. Social control and reproduction activities are generally carried out by highly visible, politically exposed public agencies ("main line" city departments, boards of education, and so on) whose programs are funded principally by local property taxation, while those activities supporting accumulation and reproduction are more often performed through insulated, politically invisible government agencies and special authorities (Friedland, Piven, and Alford, 1977). Thus, in an era of cutbacks those services with redistributive impacts become highly politicized and are often slashed while "basic," infrastructural services are held sacrosanct in all but the most distressed cities.

In addition to this functional specialization, local government is also well-suited to serving the interests of capital in conflicts over public service levels by virtue of its geographic fragmentation. Given the way in which classes and firms are sorted into jurisdictions within metropolitan areas, production-supporting and reproduction-supporting state functions are often performed to a large degree by different units of government. This means that state activity can be reduced across the full range of government services in those jurisdictions which have become less crucial to accumulation processes without endangering the production-support function of the state; higher levels of services can be maintained in other, more fiscally sound jurisdictions without burdening the accumulation process with heavy costs for the state's social control and reproductive activities. The tension between state activity and accumulation is therefore never crystalized as a class conflict, but appears as a conflict between the residents of declining cities and their own governments. This may have the effect of politicizing service issues, but it also contributes to the continued depoliticization of the economic structure and urban spatial organization which created the stress on public services in the first place.

The most crucial politics of urban service delivery may, then, be the politics surrounding the processes of local government fragmentation, for they largely set the fiscal capacity of local governments for providing services, and thereby delimit the legislative and administrative distributional decisions examined earlier in this chapter. If we are to understand the politics of urban services we must explore in greater depth the implications of the geopolitical organization of our cities for both service delivery patterns and the overall functioning of local political economies. This will require comparative research.

Critical scholars have extensively documented the intimate relationship that exists between capitalist economics and the political organization of and distribution of population in urban space (Gordon, 1976; Harvey, 1978; Mollenkopf, 1978; and Walker, 1981). To the extent that population distributions reflect the structure of the economic system, it is capitalism that gives political meaning to the geographic distribution of services. The specific political significance of geographic distributions, however, is extensively shaped by a nation's political institutions (Fainstein and Fainstein, 1978; Skocpol, 1980).

The preceding discussion of service distributions in the United States stands in contrast to the patterns observed in many other capitalist nations (Newton, 1981). The differences may be largely attributable to differences in the political institutions through which services are financed and controlled. In his study of public service production in Australia, a nation which has centralized service financing to the regional level, Andrew Parkin (1980) found substantially greater uniformity of services across jurisdictions and less political fragmentation in metropolitan areas than in the United States. He attributed much of the difference to the distinctive funding and management arrangements that mark the two nations.

It is reasonable to hypothesize that the economic and political incentives to class separation set up by heavy reliance on local funding and control of public services in the United States are a major cause of the differences that exist between the United States and European nations in the degree of segregation of classes and land uses both within and between local jurisdictions. That segregation in the United States is a major contributor to central city fiscal crises as fiscal resources flee to suburban and exurban jurisdictions, service-dependent populations concentrate in inner cities, and central city governments face spiraling costs in maintaining their huge capital investments (James, 1981).

From this perspective we can predict that the Reagan administration's efforts to further decentralize the funding of public services by dramatic reductions in the federal government's aid to cities will only encourage further, and perhaps make more severe, urban fiscal crises. The locus of the crises can be expected to shift over time. For example, when central city land values fall far enough to offset the tax cost of investing in inner city jurisdictions, and central city governments have trimmed their social consumption and expense outlays under the discipline of fiscal stress (Shefter, 1980), capital begins to flow once again into the central city, as witnessed by the economic recovery of Manhattan (Fainstein and Fainstein, this volume). Unless there is overall economic growth, however, such reverse capital flows are likely to leave other (probably suburban) jurisdictions fiscally stressed.

Local government fiscal crises, then, should be seen, not as preludes to some apocalyptic decline of the national state, but as the political reflection of fluctuations in the economic sector. Rather than collapse we might expect continuing interregional and intrametropolitian shifts of capital which carry with them shifts in the fortunes of local governments and their citizens (Gorz, 1977). We might expect this process to be intensified in the United States to the extent that the

Reagan administration succeeds in further decentralizing service financing and control by reducing the federal role in service provision.

This result, however, may be mitigated by another aspect of supply-side policy. If critical theorists are correct about the dynamics of fiscal crises, the Republicans, ironically, may have the correct prescription for easing fiscal stress. If they can engender a genuine contraction of the public sector, they may succeed in reinvigorating private accumulation sufficiently to increase government revenues indirectly and ease fiscal stress both by reducing the outlays and expanding the income of local governments across the board.

Clearly this would be a temporary victory, for it would come at the cost of much social hardship that would be likely to give rise to threats to the economic order that would call for renewed investment in social consumption and social expense activities by government. Just as the political ferment and public sector growth of the 1960s can be seen as a response to conditions created by the relative overcommitment to social investment outlays in the 1950s, the 1990s might be expected to see local governments exhaust any fiscal surplus they gain as a result of retrenchment in the 1980s as they struggle to contain the conflict and reverse the decline of labor reproduction brought about by that very retrenchment.

Budget cutting may treat the symptoms of capitalist crisis, but it does not remove the contradictions from which they are born. Hill (1978: 225-226) summarizes the relationship between urban development and capitalism in the United States as follows:

> Capital accumulation requires urbanization but urbanization requires investment, consumption, and expense outlays that market exchange cannot handle. This dramatically increases the role of state enterprise in the economy. But the structure of state production, particularly at the level of local government, is only partially complementary to private accumulation. Federalism, the fragmented system of local governments, and the structure of city-government production increasingly prevent . . . cities from accommodating themselves to the requisites of monopoly capital accumulation or the social needs of . . . residents. The result is fiscal crisis and intensifying political struggle.

While patterns in the public production of services clearly have distinctly political determinants (as Shefter [1980] demonstrates), they must also be seen as products of the functioning of a larger political economy if we are to devise effective mechanisms for their control. This understanding, however, is not advanced by dogmatic devotion to the conceptual apparatus of any particular school of thought. Eric Nordlinger's (1981) recent work should sensitize us to the subtle complexity of the challenge of specifying the role of the state in capitalist systems, and we should look to those works that bridge the gap that too often exists between critical theory and behavioral research and conceptualization of points of departure (see for example, Fainstein and Fainstein, 1978; Mollenkopf, 1979; Peterson, 1981; Shefter, 1980; and Skocpol, 1980.)

NOTE

1. Marxist scholars have been deeply divided over the nature and functions of the state in capitalist societies and over the conceptual role of the state in critical theory. (For reviews of this debate, see Dear and Clark [1978]; Gold, Lo, and Wright [1975]; Holloway and Picciotto [1978]; and Jessop [1977].) While the division has certainly not been exclusively between Europeans and Americans, there have been major differences in the approach taken on the two continents. European scholars have stressed those aspects of state action that serve to support essential institutions of capitalism, viewing the state primarily as a body that insures the reproduction of capitalist forms (Castells, 1978; Pickvance, 1976). American structuralists, by comparison, have stressed the redistributive functions of state action, depicting the relationship between the state and capitalists forms as more problematic, and the state as a catalyst to politicization of economic conflict (Katznelson, 1976; Mollenkopf 1978). This difference of opinion is possible because Marx never fully elaborated his theory of the state, but it has probably generally followed continental lines largely because of the different stages of development of the welfare state in Europe and the United States. Europeans focus more heavily on those products of state action that are collectively consumed while Americans stress the divisions reflected in, and created by, state action that distributes benefits among groups (Gough, 1979). In the analysis which follows, I make no effort to resolve these differences. I do hope to demonstrate the applicability of contemporary critical theory to the study of public service production by making only those assumptions about the nature of the state on which there is widespread agreement among critical theorists. (For an important effort to harmonize U.S. and European scholarship on this issue, see Research Planning Group, [1978].)

REFERENCES

Advisory Commission on Intergovernmental Relations [ACIR] (1973) City Financial Emergencies. Washington, DC: U.S. Government Printing Office.
ATUNES, G. and J. PLUMLEE (1977) "The distribution of urban public service: ethnicity, socioeconomic status and bureaucracy as determinants of the quality of neighborhood streets." Urban Affairs Quarterly 12: 312-332.
ASHFORD, D.E. (1980) "Centre-local financial exchange in the welfare state," pp. 204-220 in D.E. Ashford (ed.) Financing Urban Government in the Welfare State. New York: St. Martin's Press.

BENNETT, R.J. (1980) The Geography of Public Finance. New York: Methuen.

BERRY, B.J.L. (1973) Growth Centers in the American Urban System, Vol. 1. Cambridge, MA: Ballinger.

BOOTS, A. et al. (1972) Inequality in Local Government Services. Washington, DC: Urban Institute.

BOWLES, S. (1975) "Unequal education and the reproduction of the social division of labor," pp. 38-66 in M. Carnoy (ed.) Schooling in a Corporate Society. New York: David McKay.

BROWN, M.K. and S.P. ERIE (1981) "Blacks and the legacy of the great society: the economic and political impact of federal social policy." Public Policy 29: 299-330.

CASTELLS, M. (1978) City, Class, and Power. London: Macmillan.

——— (1977) "Towards a political urban sociology," pp. 2-88 in M. Harloe (ed.) Captive Cities. New York: John Wiley.

COX, K.R. and F.Z. NARTOWICZ (1980) "Jurisdictional fragmentation in the American metropolis: alternate perspectives." International Journal of Urban and Regional Research 4: 196-211.

DAVID, S.M. and P. KANTOR (1979) "Political theory and transformations in urban budgetary arenas: the case of New York City," pp. 183-220 in D.R. Marshall (ed.) Urban Policy Making. Beverly Hills, CA: Sage.

DANIELSON, M.N. (1976) The Politics of Exclusion. New York: Columbia University Press.

DEAR, M. and G. CLARK (1978) "The state and geographic process: a critical review." Environment and Planning. 10: 173-183.

FAINSTEIN, N.I. and S.S. FAINSTEIN (1974) Urban Political Movements. Englewood Cliffs, NJ: Prentice-Hall.

FAINSTEIN, S.S. and N.I. FAINSTEIN (1978) "National policy and urban development." Social Problems 26: 125-146.

FRIEDLAND, R.F., F. PIVEN, and R.R. ALFORD (1977) "Political conflict, urban structure and the fiscal crisis." International Journal of Urban and Regional Research 1: 447-472.

GILDER, G. (1981) Wealth and Poverty. New York: Basic Books.

GOLD, D.A., C.Y.H. LO, and E.O. WRIGHT (1975) "Recent developments in Marxist theories of the capitalist state." Monthly Review 27: 29-43.

GOLD, S.D. (1974) "The distribution of urban government services in theory and practice: the case of recreation in Detroit." Public Finance Quarterly 2: 107-130.

GORDON, D. (1976) "Capitalism and the roots of the urban crisis," pp. 82-112 in R. E. Alcaly and D. Mermelstein (eds.) The Fiscal Crisis of American Cities. New York: Vintage Books.

GORZ, A. (1977) "The reproduction of labor power: the model of consumption," pp. 27-39 in J. Cowley et al. Community or Class Struggle? London: Stage 1.

GOUGH, (1979) The Political Economy of the Welfare State. London: Macmillan.

HARVEY, D. (1981) "The urban process under capitalism: a framework for analysis," pp. 91-121 in M. Dear and A.J. Scott (eds.) Urbanization and Urban Planning in Capitalist Society. New York: Methuen.

——— (1978) "The urban process under capitalism: a framework for analysis." International Journal of Urban and Regional Research 2: 101-131.

HILL, R.C. (1978) "Fiscal collapse and political struggle in decaying central cities in the United States," pp. 213-240 in W.K. Tabb and L. Sawers (eds.) Marxism and the Metropolis. New York: Oxford University Press.

————— (1977) "Two divergent theories of the state." International Journal of Urban and Regional Research 1: 37-43.

————— (1974) "Separate and unequal: governmental inequality in the metropolis." American Political Science Review 68: 1557-1568.

HIRSCH, J. (1981) "The apparatus of the state, the reproduction of capital and urban conflicts," pp. 593-607 in M. Dear and A. J. Scott (eds.) Urbanization and Urban Planning in Capitalist Society. New York: Methuen.

HOLLOWAY, J. and S. PICCIOTTO [eds.] (1978) State and Capital. London: Edward Arnold.

JAMES, F. J. (1981) "Economic distress in central cities," pp. 19-49 in R. W. Burchell and D. Listokin (eds.) Cities Under Stress. New Brunswick, NJ: Rutgers University, Center for Urban Policy Research.

JESSOP, R. (1977) "Recent theories of the capitalist state." Cambridge Journal of Economics. 1: 353-374.

JONES, B. D. (1980) Service Delivery in the City. New York: Longman.

JUDD, D. (1979) The Politics of American Cities. Boston: Little, Brown.

KATZNELSON, I. (1976) "The crisis of the capitalist city: urban politics and social control," pp. 214-229 in W. Hawley and M. Lipsky (eds.) Theoretical Perspectives on Urban Politics. Englewood Cliffs, NJ: Prentice-Hall.

LAFFER, A. B. and SEYMOUR, J. (1979) The Economics of the Tax Revolt. New York: Harcourt, Brace, Jovanovich.

LEVY, F. S., J. MELTSNER, and A. WILDAVSKY (1974) Urban Outcomes. Berkeley: University of California Press.

LINEBERRY, R. L. (1977) Equality and Urban Policy. Beverly Hills: Sage.

MICHELSON, S. (1975) "The political economy of public school finance," pp. 194-228 in M. Carnoy (ed.) Schooling in a Corporate Society. New York: David McKay.

MOLLENKOPF, J. H. (1981) "Paths toward the post industrial service city: the northeast and the southwest," pp. 77-112 in R. W. Burchell and D. Listokin (eds.) Cities Under Stress. New Brunswick, NJ: Rutgers University, Center for Urban Policy Research.

————— (1979) "Untangling the logics of urban service bureaucracies: the strange case of the San Francisco municipal railway." International Journal of Health Services 9: 255-268.

————— (1978) "The postwar politics of urban development," pp. 117-152 in W. K. Tabb and L. Sawers (eds.) Marxism and the Metropolis. New York: Oxford University Press.

MOLOTCH, H. (1980) "The city as a growth machine: toward a political economy of place," pp. 129-150 in H. Hahn and C. Levine (eds.) Urban Politics. New York: Longman.

MORRIS, R. S. (1980) Bum Rap on America's Cities. Englewood Cliffs, NJ: Prentice-Hall.

MULLER, T. (1981) "Changing expenditures and service demand patterns of stressed cities," pp. 226-300 in R. W. Burchell and D. Listokin (eds.) Cities Under Stress. New Brunswick, NJ: Rutgers University, Center for Policy Research.

MUSIL, J. (1980) Urbanization in Socialist Countries. New York: M. E. Sharpe.

NEWTON, K. [ed.] (1981) Urban Political Economy. New York: St. Martin's.

————— (1976) "Feeble governments and private power: urban politics and policies in the United States," pp. 37-58 in L. H. Masotti and R. L. Lineberry (eds.) The New Urban Politics. Cambridge, MA: Ballinger.

NIVOLA, P. S. (1979) The Urban Service Problem. Lexington, MA: D. C. Heath.

NORDLINGER, E. A. (1981) On the Autonomy of the Democratic State. Cambridge MA: Harvard University Press.

O'CONNOR, J. (1981) "Accumulation crisis: the problem and its setting." Contemporary Crisis. 5: 109-125.

——— (1973) The Fiscal Crisis of the State. New York: St. Martin's.

PARKIN, A. (1980) "Centralization and urban services: the Australian experience." Policy Studies Journal 9: 1059-1065.

PEIRCE, N. R. and J. HAGSTROM (1981) "Inner city in three countries," pp. 141-155 in G. G. Schwartz (ed.) Advanced Industrialization and the Inner Cities. Lexington, MA: D. C. Heath.

PETERSON, P. E. (1981) City Limits. Chicago: University of Chicago Press.

PICKVANCE, C. G. (1976) Historical materialist approaches to urban sociology," pp. 119-146 in C. G. Pickvance (ed.) Urban Sociology. London: Tavistock Publications.

POOLE, R. W., Jr. (1980) Cutting Back City Hall. New York: Universe Books.

REISCHAUER, R. (1981) "The economy and the federal budget in the 1980's: Implications for the State and Local Sector," pp. 13-38 in R. Bahl (ed.) Urban Government Finance, Urban Affairs Annual Reviews. Vol. 20. Beverly Hills: Sage.

Research Planning Group on Urban Social Services (1978) "The political management of the urban fiscal crisis." Comparative Urban Research, 5: 71-84.

RICH, R. C. [ed.] (1981) The Politics of Urban Service Distribution. Lexington, MA: D. C. Heath.

——— (1980) "The complex web of urban governance, gossamer or iron?" American Behavioral Scientist 24: 277-298.

——— (1979) "Distribution of services: studying the products of urban policy making," pp. 237-260 in D. R. Marshall (ed.) Urban Policy Making. Beverly Hills: Sage.

ROMANOS, M. [ed.] (1979) Western European Cities in Crisis. Lexington, MA: D. C. Heath.

SANGER, M. B. (1981) "Are academic models of urban service distribution relevant to public policy? lessons from New York." Policy Studies Journal 9: 1011-1020.

SBRAGIA, A. (1981) "Cities, capital, and banks: the politics of debt in the United States, United Kingdom, and France," pp. 200-220 in K. Newton (ed.) Urban Political Economy. New York: St. Martin's.

SHEFTER, M. (1980) "New York City's fiscal crisis: the politics of inflation and retrenchment," pp. 71-94 in C. H. Levine (ed.) Managing Fiscal Stress. Chatham, NJ: Chatham House Publishers.

SKOCPOL, T. (1980) "Political response to capitalist crisis: neo-Marxist theories of the state and the case of the new deal." Politics and Society 10: 155-201.

SPERO, S. and J. M. CAPOZZOLA (1973) The Urban Community and Its Unionized Bureaucracies. New York: Dunellen.

STANLEY, D. (1977) "The ambiguous role of the urban public employee," pp. 23-35 in C. Levine (ed.) Managing Human Resources. Beverly Hills: Sage.

STONE, C. (1980) "Systemic power in community decision making: a restatement of stratification theory." American Political Science Review 74: 978-990.

SZELENYI, I. (1981) "The relative autonomy of the state or state mode of production?" pp. 555-591 in M. Dear and A. J. Scott (eds.) Urbanization and Urban Planning in Capitalist Society. New York: Methuen.

THOMPSON, E. P. (1978) The Poverty of Theory and Other Essays. London: Merlin Press.

Twentieth Century Fund Task Force on Municipal Bond Credit Ratings (1974) The Rating Game. New York: Twentieth Century Fund.

U. S. Bureau of the Census (1979) Government Finances and Employment at a Glance. U.S. Washington, DC: U.S. Government Printing Office.

WALKER, R. A. (1981) "A theory of suburbanization: capitalism and the construction of urban space in the United States," pp. 383-429 in M. Dear and A. J. Scott (eds.) Urbanization and Urban Planning in Capitalist Society. New York: Methuen.

WATKINS, A. J. (1980) The Practice of Urban Economics. Beverly Hills: Sage.

WEITZMAN, J. (1979) City Workers and Fiscal Crisis. New Brunswick, NJ: Rutgers University, Institute of Management and Labor Relations.

WOLCH, J. R. (1981) "The location of service dependent households in urban areas." Economic Geography 57: 52-67.

YATES, D. (1979) "The mayor's eight-ring circus: the shape of urban politics in its evolving policy arenas," pp. 41-70 in D. R. Marshall (ed.) Urban Policy Making. Beverly Hills: Sage.

ZACK, A. M. (1973) "Meeting the rising cost of public sector settlements." Monthly Labor Review (May).

Part V

Urban Social Movements:

Culture, Class, and Consciousness

10

Race and Schooling:
Reflections on the Social Bases
of Urban Movements

IRA KATZNELSON

KATHLEEN GILLE

MARGARET WEIR

☐ MOST PEOPLE WHO are oppressed do not revolt. When they do, they undermine the values and practices which make oppression appear natural, or even the fault of the oppressed. The risks they run demonstrate "the infusion of iron into the human soul to give it the power to judge and to act" (Moore, 1978: 182). Courage to challenge the prevailing culture is not only a matter of individual psychology but of collective mobilization as well. Revolts depend on group formation — the transformation of groups of people sharing attributes of geography, sex, kinship, religion, language, race, or class into groups in action.

The urban social movements which have been an important element of political conflict throughout the West in the past quarter century have been based on various group characteristics. Most have had class as the main, or as a leading, element. Even where urban movements have been wholly divorced from the concerns of workers

AUTHORS' NOTE: This chapter has been prepared as part of the National Opinion Research Center's project on "Urban School Organization and the American Working Class: An Historical Analysis," funded by the National Institute of Education.

215

as labor, they have articulated demands in a rhetoric of class, typically as members of a working-class community, and in so doing, have broadened the meaning of class and class struggle.

Not so in the United States, where race, not class, has defined the most important insurgent movements. This familiar observation provides our point of departure and our central empirical question. The obvious but unsatisfactory general answer is that blacks are more oppressed by whites than workers by the bourgeoisie. Yet the most oppressed need not be the most insurgent; usually they are not. Further, race and class need not be mutually exclusive bases of revolt. The absence of class is as striking as the presence of race at the center of the American urban movements of the 1960s. In this chapter, we take up a critical, perhaps *the* critical, instance of urban conflict in the recent past: black school movements.

If, in so doing, our first question is to explain the distinctive social basis of American urban movements in the period of the urban crisis, our second, more broad, question is part substantive, part methodological. How are we to account for the specific social bases of urban movements in various settings? This issue is especially pressing if we wish to take seriously the similarities of urban movements in different places even as we wish to understand their important variations.

CHARACTERISTICS OF URBAN MOVEMENTS

Urban movements have shared characteristics across national boundaries. They have organized outside the workplace. They have been concerned with the delivery of services by government and with the impact of housing, transport, education, and social services on the spatial and social structures of the city. They have linked the condition of specific classes and groups to the provision of collective goods and services, and in so doing, have created new ways of understanding poverty and inequality.

American movements of black city residents, including school movements demanding equal access and equal provision of public education, have shared these characteristics. Indeed, internal colonial analysis, the nationalist-inspired ideology of some of the more militant groups, provided a rather pristine crystallization of the main traits of urban movements by identifying the relationship between blacks in specific city ghettoes and service bureaucracies as the key element in a system of domination. Black struggles, like urban struggles elsewhere, were struggles against the providers of government largesse.

The similarities between the various western urban movements have provided the empirical underpinnings in the past decade for the development of important theoretical critiques of liberal urban analysis, and for macro-level attempts to construct alternatives that directly connect patterns of capitalist and urban development. This intellectual (and political) trend may be dated from the publication in 1972 of Manuel Castells' enormously influential *La Question urbaine*. Primarily informed by a structuralist reading of Marx, this school has stressed the place of the city in advanced capitalism, the role of the state in shaping the spatial organization of the city, its patterns of economic transaction, and the importance of urban movements. By incorporating space (its production and consumption) and state policies (their production and consumption) into urban analysis, this work has significantly expanded the analytical power and political pertinence of studies of the capitalist city, its crises, and contradictions (Castells, 1977; see also Pickvance, 1976; Harloe, 1977).

This intellectual tradition has accorded urban movements a new legitimacy within Marxism. Urban struggles are viewed mainly as products of economic, political, and spatial developments, which, ultimately, are driven by the demands of capitalist expansion. An urban movement is defined not only by the larger structures which give it life, but also by its effects on the existing urban system, and ultimately, on the reproduction of capitalism. For Castells, a *movement* qualifies for the term, and thus becomes a candidate for structurally informed urban analysis, "to the extent that it attempts to question the logic of the urban system" (Castells, 1978: 132). The behavioral hallmark of such struggles is that they raise demands that cannot be resolved or absorbed by regular political procedures or substantive solutions. In doing so, they bypass ordinary patterns of political participation.

BLACK SCHOOL MOVEMENTS

By these criteria the black school movements of the United States in the 1960s fall within the ambit of structuralist scrutiny. After all, the central feature of the challenges they posed was their unruly questioning of school politics as usual. Black demands upset decision-making patterns that had been elaborated over the years by local school districts. These demands for integration, community control, and equal provision proved impossible to resolve by local authorities

using ordinary methods of resolving disputes. The most important
indicator of the inappropriateness of regular politics was the judi-
cialization of school issues.

These features of the interactions between blacks and the schools
were impressed on us in the course of a study we have been conduct-
ing of the politics of schooling and the working class in Atlanta,
Chicago, and San Francisco from the mid-nineteenth century to the
1970s. In spite of local variations, we observed a similar politics of
race and education in each city. Black demands, even when initially
raised in modest form, proved threatening to a wide array of whites: to
neighborhood residents who feared racial integration and its expected
consequences; to politicians who feared the loss of traditional constituents as a result of population change; to school officials who
sought to preserve time-worn patterns of organization, definitions of
school issues, and personal political ties. With varying degrees of
articulate consciousness, whites also understood that the black attack
on the exclusionary policies of the public schools was part of a
broader effort to achieve social, economic, and political equality.
Thus, the targets of black demands felt all the more imperiled because, contrary to the fragmentary, issue-specific character of school
politics as usual, black school struggles were linked to a wider, more
global set of concerns. The importance given to education by most
Americans as the key to opportunity (a view shared by blacks who
placed a high priority on schooling) raised the stakes.

The result was a dialectic of white resistance and an intensification
of black activity which became increasingly unruly (and which, in
some cities, included a change from a focus on integration to community control). Ordinary patterns of dispute resolution were utterly
incapable of managing conflict with this content, meaning, and intensity. The shift of the arena of school politics from school boards,
bureaucracies, city halls, and city councils to the courts is both an
indicator of the disruptive capacity of black school movements and of
the search for a "nonpolitical" forum to manage school politics and
restore order. As both Robert Dahl (1957) and Richard Funston(1975)
have demonstrated in their work on the Supreme Court, courts tend
to take a more activist role when the legitimacy of public policies is
under severe strain, and when conditions of crisis prevail.

It was the Supreme Court itself, of course, in the landmark *Brown
v. Board of Education* (1954) that made possible the reopening of
fundamental questions of race and schooling. In Atlanta, where
segregation was mandated by law, *Brown* shifted black demands from
pleas for improved facilities within Jim Crow to calls for the total
abolition of the dual school system. Although *Brown* did not immediately precipitate challenges to school policy in the North—it did

not clearly apply to cities like Chicago and San Francisco where school segregation resulted mainly from residential patterns— the demonstration effects of the southern civil rights movement stimulated by *Brown* helped produce northern school movements which tried to define the issues of segregation, access, and the terms of access to education in ways appropriate to northern cities.

As in Atlanta, fundamental assumptions about race relations were questioned. In the southern city the challenge was to a policy of overt separatism; in the North, to one officially color blind—but in both to the ideological assumptions and basic practices of local politicians and school officials. Even when the influence of Black Power ideology created a rupture between integrationists and separatists in the black community, all the movement actors understood that dealing with racial issues required fundamental departures from school politics as usual. In Chicago and San Francisco, as much as in Atlanta, even ordinary issues such as curriculum, personnel, governance, and school location became increasingly contentious as they were tinged with racial overtones.

Local school districts lacked the capacity or, in many instances, the will to respond to calls for such profound changes, and extraordinary tactics were required before the full force of black demands was acknowledged. In Atlanta, when petitions for integration to the local school board brought little action, black leaders went directly to the federal courts and filed suit in 1958. In spite of a ruling favorable to the cause of integration, implementation was slow, in part because Georgia state law initially prohibited integrated schools. Faced with this contradiction between federal and state decrees, local officials chose to abide by the state statute and did little to comply with the first federal court ruling until the integration of the University of Georgia led to the repeal of Jim Crow. Still, local administrative resistance and the fear of white backlash kept Atlanta's integration program small throughout the early 1960s. Blacks were recurrently drawn back to the courts to press for more comprehensive plans. Eventually, an out-of-court compromise was reached between the NAACP and the local school district. In 1973, the black leadership virtually dropped its demand for integration of the public schools in exchange for meaningful participation in the administration of the school system. Thus, despite early court precedents favoring major alterations in school racial policies, a full-scale crisis was averted only through methods of elite mediation and in the context of federal litigation—arenas far removed from the traditional sites of school decision making.

In Chicago and San Francisco, the absence of early court precedents dealing with de facto segregation and the different political heritage of each city led to greater group mobilization and more virulent conflicts. Blacks took their demands to the streets, especially in the early 1960s when boycotts, picketing, sit-ins, marches, and other confrontational tactics were greeted by the adamant opposition of tradition-minded educators. As the depth of black grievances became apparent in the mid-1960s, school officials sought to placate some of their demands. Massive studies of racial problems were commissioned by local officials. In San Francisco, newly appointed prointegration members of the school board and a new superintendent promoted a plan to use federal funds for integrated educational complexes requiring minimal amounts of student busing. The militant opposition of white neighborhood groups, encouraged by the mayor, undermined these plans and the local stalemate was resolved only when civil rights groups sought federal court action. San Francisco became the first northern city to undergo city-wide busing to achieve racial balance in its schools when in 1971 a federal district court decided that recent U.S. Supreme Court rulings applied to de facto segregated cities.

In Chicago, local civil rights groups called on federal officials to withhold funds from the school district as provided for in Title VI of the 1964 Civil Rights Act. As a result, demands for change became entangled in political cross-pressures between local and federal officials, impeding progress toward integration. Even when the superintendent, who had become a national symbol of opposition to integration, resigned, the objection of white neighborhood groups to busing forestalled integration efforts. Although legal action was threatened, each time blacks turned to the courts, a solution short of major reorganization of the Chicago school system was worked out. By the end of the decade, Chicago's schools remained highly segregated and local school officials seemed immobilized in the face of the conflicting demands of white and black groups and the ever-present threat of federal sanctions or an unfavorable court ruling.

In each city black school movements precipitated a crisis for local government. The historical exclusion of blacks from mainstream school services and policy-making created conditions for demands that, by their nature and tactics, radically challenged the status quo. Resistance and political stalemate undermined the institutions of local school governance. As blacks sought redress from authorities beyond the local jurisdiction, federal agencies and especially the federal courts became the key conflict managers. In the long run, however, the failure of conflict resolution at the local level took its toll. Black grievances, white resistance, and the political incapacity of local school districts left racial issues unassimilated into routine school policy. Thus, today the politics of race still retains its disruptive potential, just as educational equality remains elusive.[1]

APPLICABILITY OF STRUCTURAL ANALYSIS

The structuralist school has provided a compelling intellectual perspective to make sense of these events. It demonstrates that urban blacks, like social classes elsewhere, had a distinct role in the larger political economy, a distinct social geography, and a distinct relationship to collective services, including education. Black urban movements, like others, are products of general social, economic, and political processes, and like others, tried to affect basic distributions of material, cultural, and power resources. Further, by locating the black insurgency within a larger analysis of the urban crisis in capitalist cities, the structural perspective helps us understand why the 1960s were a propitious time for revolt. In this way, the black story can be told in a manner that is neither parochial nor ad hoc.

However, there are specific features of the American case with which such an analysis, no matter how well-elaborated, cannot grapple, the clearest instance being the movements' social basis. The structuralist approach can locate black movements within a larger family, and it can provide an explanatory agenda for an analysis of the causes and effects of these movements. But it is incapable of explaining why race, not class, was so central in the United States.

Indeed, the rhetoric of structural treatments of urban movements appears to assume a class orientation. Castells, for example, talks about them in terms of "class political relations" and "class ideological relations" (Castells, 1978: 126ff). Urban struggles are seen as an aspect of class struggle. One of two interpretations is possible, each unsatisfactory. Either all urban movements are thought to be class movements or all urban struggles, whatever their basis, are automatically assimilated to the category class struggle. The first is empirically wrong; the second, by begging too many questions, is theoretically impoverished. When struggles change their historical forms they are not the same struggles. Thompson (1974: 68) cautions, "We cannot know of struggle in a formless, decontextual way." It is one thing to say that race conflicts express and work themselves out according to some features of the logic of capitalist development, but it is quite another to treat all conflicts as class antagonisms.

We think, in short, that the inappropriateness of structural urban work to our question about the social basis of American urban move-

ments reflects a more general problem. Structuralist gains have come
at a price. Travel from abstract theory to concrete cases that are more
than ad hoc illustrations of the power of the general theory has proved
difficult. This chapter, concerned as it is with racial conflict about
public education in the United States, provides an occasion for considering by example how to traverse the distance between theory and
cases. Too often this journey is made by presenting descriptive case
materials merely to illustrate theoretical claims. The recent urban
literature is full of case studies, but their problem is that they do not
take the instances, and their variety, seriously enough. Rather, cases
tend to be treated as exemplars or as representative occasions. The
new urban social science, as a result, has been more interested in
common elements which allow cases to compose a family than with
systematically identifying and accounting for variations between
members of the family (see Skocpol and Somers, 1980).

These are different, if complementary, tasks, and all the hortatory
pleas for "specificity" will not substitute for macrosocial comparative
analysis. We must treat the call for specificity not as one for illustra-
tion or supportive detail, but as a plea for developing tools capable of
analyzing differences systematically. This appeal is not just academic,
even in the best sense, for without this kind of specificity it is hard to
see how strategic political reasoning can occur. No one lives or
experiences capitalism in general.

COMPARISONS

Our empirical focus implies two comparisons: between the politics of schooling for whites and blacks in the United States; and
between the American pattern and other national situations at comparable structural moments. Each is directed at understanding the
relationship between particular groups and a particular area of public
policy, in this case education. In cross-national perspective both the
relevant policies and groups vary from society to society, even if they
share in broad constraints dictated by the logic of the mode of production. Thus, to say that a state in capitalism is a class state is to say
rather little unless this assertion is elaborated to distinguish between
the permanent and the mutable in the relationship between state and
capital. Within the conditioning limits identified by Marx (and, we
would add, the organizational limits identified by Weber) states
authoritatively act to tax, conscript, and make public policy.
Capitalist states, as we all know, vary significantly in how they carry
out these activities.

The character of groups and classes sharing dispositions and capable of organization and action also varies from society to society. Even if we limit our gaze to questions of working class formation, it is obvious that no national working class is exactly like that of any other. It is our claim that it is in the study of the connections between these variations in policy and class and group formation that we can account for the distinctive histories of various urban movements, including American school movements based on race. At issue is the specific set of linkages that have joined various groups and classes to specific public services.

The historical development of modern states and the expansion of their reach into new areas of public policy are aspects of the process of state-building, in which the state becomes increasingly differentiated and autonomous from other organizations, and in which the state controls, restricts, or eliminates competitors. This general process — and particular instances of it such as the creation of public school systems — entails the demarcation of increasingly clear borders between the state and civil society. These boundaries are places of transactions between groups in civil society and the state.

Just as the rate at which new domains of public policy are created has varied a great deal from country to country and between policy areas within countries, so the links forged between the new policy areas and specific groups have been uneven: Groups differ with respect to the *timing* of their inclusion; and, once the policy includes them, they differ with regard to the *terms* of their incorporation. Further, whether they are involved with the state's policy or not, groups have different *capacities* to affect the state's policies or to provide alternatives to them.

Variations in timing, terms of incorporation, and group capacities establish distinctive types of linkage between groups and the policy arenas which have logical consequences for the process of group formation, for the character of policy demands the group will make, and for the content of the policy arena (see Tilly, 1975).

Consider the relationship of exclusion and resources in nineteenth-century England. Before the passage of the Education Act of 1870, no national system of primary school provision which assured schooling to working-class children existed. Workers were excluded, but unlike the dominant classes, they were incapable of providing satisfactory "private" solutions (though they made many attempts to do so). As a consequence, their exclusion helped shape their formation *as a class in action* to demand state education. In the United States, by contrast, when modern school systems were first established in the years before the Civil War, workers and their families did not have to struggle in an equivalent way for mass schooling (see Katznelson, Gille, and Weir, 1981). Their inclusion was a part of what was understood by a wide array of social groups and classes to be the minimum state responsibility for free mass education. This early and relatively painless inclusion affected the basis and character of working-class school demands. They were made not by a class demanding inclusion but by groups that varied from issue to issue. Only on occasion were school concerns articulated and understood in class ways.

Once included in a policy domain of the state, groups are linked to it in one of four broad ways that structure transactions across the boundary of state and civil society. First, the group may be incorporated fully and equally, and may possess the capacity to affect the contours of the policy. Second, the group may be incorporated fully and equally in the policy arena, but be relatively incapable to affect it. Third, the group may be incorporated in partial or structurally subordinate ways, yet it may possess the capacity, at least at some moments, to influence the state's policy activities. And, fourth, the group may be structurally subordinate in the policy arena, and without resources to affect what the state does.

After the 1870 Act ushered in mass schooling in England, the working class was incorporated in a manner that reproduced class inequalities in explicit ways, especially but not only at the level of secondary education (see Williams, 1961: 125-165). Workers continued to pose curricular and other demands in class terms through the institutions of an organized working class which, in part, had been formed in opposition to the earlier exclusion of working-class families from access to mass schooling. Working-class capacity to affect school policy was tied intimately from the outset to the larger battles fought by the labor movement. The class basis of demands by school movements in England in the 1960s for the provision of comprehensive secondary schooling is at least partially comprehensible in these terms.

In the United States where the working class had been incorporated fully and equally without similar struggle, the politics of education concerned issues that assumed inclusion in a system of elementary schooling and the provision of a minimum standard of instruction. Because white (male) workers in the United States, unlike their English counterparts, were also citizens in the full sense of possessing the right to vote, and because they were mobilized as voters into the political process by mass political parties they possessed the ability to influence the contours of urban school systems. We found that workers in Atlanta, Chicago, and San Francisco were concerned with three broad areas of schooling questions: (1) Matters of schooling tied directly to the occupational order and the workplace. Here American workers appeared in the school arena as labor. (2) Cultural and communal questions, such as language instruction. Here workers appeared as ethnics. And (3) issues of organizational maintenance and development, including matters of governance, finance, and the territorial organization of the schools. Here workers appeared as taxpayers, ethnics, and as supporters of this or that political party.

One important consequence of the pattern of linkage between white American workers and the schools was the fragmentation of their demands into concerns about the work and off-work dimensions of their lives. Each sphere was considered in its own terms. Schools, which span the divide between work and community, did not contribute to bridging the divisions between the working class as labor and the working class as ethnic residents of specific city neighborhoods.

The American pattern assumed that the most basic issue of school politics — that of access to mass schooling — had been resolved once and for all; and it presumed that working people were incorporated fully and equally with a capacity based on political participation to affect school policy. These patterns of linkage eliminated the questions that were at the heart not only of the English but of most European battles about schools in the nineteenth, and in many cases, in the twentieth centuries. For blacks, however, the situation was conspicuously different. Immediately after the Civil War they were confronted with attempts to exclude them from mass schooling; and, after they were included, they were linked to the various urban school systems in structurally subordinate ways with very limited capacities to influence what the schools did. Even where blacks were concerned with other issues of school politics, they were at the same moment concerned with securing and maintaining a tenuous access to education and with questions of de jure or de facto segregation.

Nowhere were these patterns more pronounced than in the American South. Rigid prohibitions against racial mixing dictated that blacks would be confined to their own schools; white monopoly of local political institutions meant that blacks would have limited — and diminishing — capacity to improve the schools allotted them, much less challenge their subordinate status within that system. But because it identified education as central to the larger effort for full and equal membership in American society, the black community tirelessly searched for new ways to influence school policy. The struggle for schooling was central in fashioning blacks as a group with particular political demands and a distinct history of political practice centered around educational issues.

Faced with a constitution that granted political rights and acknowledged the state's responsibility for educating blacks, outright exclusion was not an option when Atlanta's newly formed board of education met in 1869 to design a public school system. Yet, to obtain even the legal minimum, Atlanta's black community had to place organized and persistent pressure on public officials. Once schools had been established, the board of education's continuing neglect of black educational needs handed blacks an issue around which to organize. Early provisions for black elementary schooling were markedly inferior to those of whites. Black schools were grossly overcrowded. New facilities were opened only after considerable prodding by black leaders. Moreover, no secondary education was available to blacks, and on this issue the board of education refused to yield until the 1920s.

Although blacks were active in pressuring school officials and in mobilizing their community, they formulated demands and tactics that did not fundamentally challenge the Jim Crow framework of public education in the South. In the earliest efforts to obtain better schools, black spokesmen were solicitous of white officials, persistently requesting public funds but at the same time accepting inferior facilities for fear of being denied schooling altogether. For instance, ministers who led the initial movement for schools offered their churches for sale or rent to the school board, which was reluctant to construct new buildings for blacks.

On two occasions in the nineteenth century — 1888 and 1891 — Atlanta's white electorate split into opposing camps, each attempting to lure black voters. Black leaders were circumspect in their use of such infrequent opportunities, opting for firm promises of schools rather than pressing for representation on the school board and other benefits offered by one of the competing parties. The need for concrete improvements was so great and the possibilities of successfully challenging the basic structure so remote that black leaders defined the prudent course as the most fruitful.

These episodes were critical in the formation of a black political identity and practice. The strategy of white politicians implicitly defined blacks as a group with its own distinctive needs and demands. Blacks were compelled to act as a *political* group because the rare opportunity to win benefits through the use of political power depended crucially on the ability to deliver votes as promised. In both instances blacks, led by an articulate church-based middle class, bargained to obtain the facilities they had requested, further under-scoring the importance of black political cohesion.

Although the passage of the white primary in 1892 plunged blacks into political oblivion, it paradoxically served to intensify unity as blacks searched for new avenues to press their claims for better educational facilities. Now barred from selecting candidates for municipal office, blacks were still able to vote on city-wide bond issues. Three times after the turn of the century (1919, 1939, and 1940), blacks launched voter registration drives and successfully voted down bond issues which did not allocate sufficient funds for their schools. This tactic required a high degree of organizational activity and group solidarity. Moreover, the victories on the bond issues encouraged Atlanta's black community to strike a more activist pose in confrontations with the city's white power structure. Many of the organizations that had participated in the bond campaigns, particu-larly the NAACP, were active in the 1941 struggle to equalize salaries for black and white teachers — the first instance in which black demands concerned equal benefits for both races. Likewise, these organizations took the lead in extending the claims for equality and integration of educational facilities in the 1950s and 1960s.

Not until black demands were backed by the federal government and an emerging national civil rights movement, however, did they challenge the structural subordination of blacks within Atlanta's pub-lic schools. Without sources of political leverage outside the local structure of power, blacks had to play according to the rules of that structure, if they were to receive anything.

In the North, the absence of a history of strict racial segregation made more uncertain the ways blacks would be bound to school systems. In both Chicago and San Francisco, early regulations man-dating segregation were successfully challenged, incorporating blacks into the public schools by 1875. Yet, because they constituted a negligible percentage of the population, northern blacks were power-less to affect school policy. Moreover, once their numbers increased, a pattern of de facto segregation took shape, effectively recasting the black relationship to schools by forging a new structurally subordi-nate linkage. As in Atlanta, the unending struggle to secure or retain equal participation in schools was a central force in uniting blacks as a group.

In San Francisco, where blacks accounted for less than one per-cent of the population, they were initially excluded from the public schools. Only persistent petitioning persuaded the board of education to establish a "colored school" several years after white schools had opened. Facilities for black children in San Francisco were inferior to those of whites, and their existence remained precarious as a school board order arbitrarily closing two black schools in 1868 so vividly demonstrated.

Better schools constituted a chief goal of black activity from 1850-1875: Black newspapers carefully monitored educational progress; black leaders used school issues to mobilize the community. The very different position of blacks in the social and political environment of the American West opened the way for alternatives unthinkable in the South, most notably a demand for school integration.

Although the size of the black community precluded the use of electoral power, blacks did have Republican supporters in the California state legislature. The combination of strong political allies and the low salience of the issue for most whites meant that the capacity of blacks to challenge their subordinate status was much greater than their numbers would suggest. After many community meetings, petitions to the board, introduction in the Assembly of a bill proposing school integration, and an 1872 test case in the Supreme Court of California, blacks had still not achieved their goal. However, in 1875, when the San Francisco board found itself under severe financial strain, it yielded to the pressure, abolishing the "colored school" and integrating blacks into white schools. There were no incidents accompanying integration, and blacks disappeared from San Francisco school politics as a group with distinct demands until the post-World War II influx of blacks established a pattern of de facto segregation, recasting the contours of racial politics.

For Chicago's blacks a similar tale of exclusion and consequent emergence of specifically black demands can be told. Here, however, the pace of events differed. Race occupied the school agenda only for a brief period during the Civil War when a segregation law was imposed mainly at the insistence of the Irish community. The law was repealed when black parents refused to remove their children from the white schools.

As the black population of Chicago grew rapidly after the turn of the century, so did calls for formal resegregation. Whites responded to the pressure of black pupils in previously all white schools with strong and occasionally violent protest. Although school officials never seriously contemplated legally reinstituting segregation, they designed policies to prevent racial mixing, thus creating a parallel but inferior black school system more like Atlanta's than San Francisco's.

Chicago's blacks, however, were not disenfranchised and the black vote, if not properly harnessed, could disrupt the designs of white politicians. In its consolidation in the 1930s, the Democratic machine successfully eliminated this possibility by including blacks but on unequal terms. Although black politicians provided a conduit through which Democrats funneled limited amounts of patronage into the black community, they were relegated to the periphery of machine affairs. Moreover, the kinds of benefits which blacks could receive were severely circumscribed by what was acceptable to the machine's white constituents. Thus, although black claims for better schools were more successful in Chicago than in Atlanta, in neither city could blacks transform the nature of their relationship to the schools. The school issue was a volatile one for Chicago's Democratic politicians: Violent opposition by whites to integrated schools and the equally insistent and organized activities of the black community around educational issues constrained the course of action followed by school officials. The school department sought to alleviate the worst instances of overcrowding while avoiding integration of blacks with students in surrounding white neighborhoods. It made attempts to appease black demands for educational equity by redressing only the most flagrant instances of discrimination and by granting blacks representation on the school board. Although they continued to press for an end to segregated schools, blacks recognized the obstacles to this goal and opted for improvement of black facilities at the expense of integrated schools. Although their voting strength gave them enough power to extract more benefits than Atlanta's blacks, Chicago's black community did not have the leverage to compel school officials to desegregate. Thus, even in the North, where segregation was not legally mandated and where blacks did demand integration, low political capacity handicapped blacks, preventing them from breaking out of their subordinate status within school policy.[2]

FORMATION OF DIRECT ACTION MOVEMENTS

With this century-old heritage of group consciousness and organization, honed in interaction with the schools and focused on the acquisition of rights, blacks entered the 1960s. Although unintegrated into traditional urban politics, they possessed a broad and deep array of institutional resources that provided what Von Eschen, Kirk, and Pinard (1971) call the "organizational substructure of disorderly politics." Sociologists who work in the "mass society" tradition have tended to assume that organizational membership "drags people into routine politics, while discouraging participation in direct action movements." This approach fails to take into account the particular history of formation, group basis, and content of organizations and movements. "For an important class of organizations — organizations with goals unincorporated by the larger society or containing members with interests or values that are unincorporated — membership increases rather than inhibits participation" (Van Eschen et al.: 529).

Put more broadly, it is crucial to understand that the formation of a group in action is a radically conditional process. As students of collective action have well understood, people who may be persuasively joined in the same analytical categories, as those of race and class, do not necessarily or naturally share dispositions or act together to secure common ends. The identities of collective actors and the character and scale of their mobilized activities demand explanation. For collective mobilization to occur, group members must recognize they share a community of fate based on common treatment by others; they must create a shared culture of some sort, based on interactions, the construction of an interpretive language, and common practices; and they must find collaborative rather than merely individual solutions compelling. And even if all these barriers to group formation are overcome, groups still face hurdles of creating and maintaining effective organizations (see Hardin, 1982).

The complicated process of the formation of groups in action does not depend only, or perhaps even primarily, on the structure of the economy and the social order it fashions, nor on the relationship in general between the state and capital at different moments of capitalist development. Rather, group formation is affected crucially by the particular timing and character of linkages joining particular groups and arenas of state policy. In the area of schooling in the United States, the relevant linkages muted class but highlighted race as bases of collective action. The state initially defined blacks as a target of exclusion from schooling; and even after blacks became clients of school systems, the terms of their incorporation continually reproduced the group as a unit of society simultaneously economic, cultural, and political. The "organizational substructure of disorderly politics" of black urban movements was the product of this quite specific set of historical relationships. The challenges which resulted indicated that black patterns of incorporation had been very different than those of white American workers (and in some ways more like those of European working classes).

AMERICAN EXCEPTIONALISM

The story of race, class, and schooling, of course, is only one instance of the general methodological problem of dealing with national variations in patterns of class and group formation while taking seriously what Thompson (1978: esp. 229-242) calls the shared "logic of process" of all capitalist societies. Put in this context, our story is also part of a larger tale of American exceptionalism, which is con- . cerned with identifying those features of the American experience that are distinctive within the universe of the capitalist and democratic societies of Western Europe and North America. Often the term is defined by the absence in the United States of what other societies are thought to have: socialist or social democratic parties; class conscious working classes; militant trade unions — the aspects of politics evident when class and class conflict are the defining features of political life. Quite aside from the fact that the development of each capitalist society may be called exceptional, treatments of American exceptionalism rarely specify in positive terms what features of the American regime are in fact distinctive.

Over time, the United States incontestably has developed a special mix of public policies which can be characterized by the formula of high schooling and low social welfare. The story of the early founding of public schools contrasts with the late development of other social policies. Not until the New Deal did the United States begin to create a national structure of social insurance and services similar to those created decades earlier in much of Western Europe. The United States has maintained its early lead in mass education through the university level, just as it still lags in comparative rankings of welfare state expenditures, and more generally, in its governmental capacity to shape and modify market forces. Thus "high schooling" cannot be considered simply on its own terms, nor just in comparison to levels of schooling elsewhere, but as part of a more complicated package of social policies.

The United States is unusual in other ways that also affect the patterning and experience of class. The history of the U.S. working class has been characterized by a pattern of class formation marked above all by a stark divide between a consciousness and a politics of work and off-work. The American working class, as a class, has been formed almost exclusively as labor, as a collectivity oriented to matters of wages and laboring conditions at the place of work. Off work, in their residential communities, American workers have acted mainly on the basis of nonclass territorial and ethnic affiliations, and they have been organized into political life by political parties which are quite independent of the trade union movement. This pattern of divided consciousness originated in the antebellum period in the port cities of the Northeast. Workers as laborers were relatively free to join trade unions (two thirds of wage workers in New York belonged to craft unions in the 1830s), and male workers as citizens were free to vote and support political parties which, in a system of participatory federalism, tended to mobilize supporters on the basis of ward-based affiliations. In turn, once formed in the special way it was, the political capacities of the American working class were limited; the comparatively low standing of the United States in social welfare expenditures may be explained in part by this factor (Katznelson, 1981: ch. 3; Katznelson, 1978).

From this vantage point, the relatively high premium Americans have put on schooling can hardly be interpreted principally as a victory for working-class struggle. Indeed, a more compelling argument can be made that the early consensus about mass primary schooling made the formation of an English (or French, or German, or Scandinavian) type working class with a consciousness spanning work and community more difficult, because American workers, unlike their European counterparts, did not have to struggle against the state for a long period on behalf of public education. Thus, whereas in much of Europe questions of education as well as those of the vote defined the working class as a political collectivity, in the United States both kinds of struggle were absent.

For blacks, the general American policy formula of "high schooling, low social welfare" at minimum was subject to modification, and for most times and places would seem a mockery. The divisions between work and off-work have been less relevant to black than white workers because their condition has been a total one. Blacks have been treated and formed as a group transcending these divisions. For blacks, the American state was not one of participatory federalism, nor did the United States appear as a weak state. Their history with respect to schools, with regard to citizenship rights and public services more generally, was one of subjugation, exclusion, and, in many instances, violent repression. Until well into the twentieth century, blacks were not part of the trade union or party institutions of working-class life.

The black school movements of the 1960s were shaped by and invoked three main features of "American exceptionalism" — a special kind of "weak" participatory state with a distinctive mix of public policies, a working class formed only as labor, and the compelling importance of racial divisions. We have intended our retelling to show how the particular dynamics of statebuilding affect national experiences of group formation. But we also urge that the time is overdue to heed Pierre Vilar's (1973: 99) injunction that "what is more urgent" than the further refinement of structuralist theory is "the elaboration of methods of passing from theory to the analysis *of cases* (or *frameworks of action*)."

NOTES

1. A wealth of detail on the politics of school desegregation in Atlanta from the period of the *Brown* decision through the mid-1960s can be found in Huie (1967). For an analysis of further efforts to desegregate leading up to the 1973 compromise, see Research Atlanta (1973). See Ecke (1972) for a useful history of school politics in this era. Black politics in Atlanta is treated in Hornsby (1973) and Suber (1975). A comprehensive study of black school movements in Chicago is yet to be written. The best analysis of school desegregation efforts is included in Orfield (1969); Chicago is treated in chapter four as a turning point in national policy concerning the enforcement of Title VI of the 1964 Civil Rights Act. Cuban (1976) has useful information on the resistance to integration during the early 1960s when Superintendent Benjamin Willis was in office. For an analysis of the period immediately following Willis' resignation, see Coons (1968). Peterson (1976) covers racial politics in the late 1960s. Kirp (1977) is a concise yet thorough analysis of the politics of race and schooling in San Francisco with special emphasis on the role of the courts. Weiner (1973) treats the response of the school system to court-ordered desegretation from the perspective of organizational theory. A volume which is valuable for placing contemporary events in San Francisco in the context of the century-long struggle for racial equality in California is Wollenberg (1976). Some information on the politics of San Francisco's black community can be found in Wirt (1974).

2. The literature is sparse on the formative years of black education in Atlanta. Racine (1969) provides the best general overview of Atlanta's schools. The relationship of the early struggle for schooling to race relations in other arenas of city life is discussed by Rabinowitz (1978). The best summary of black politics in this era is Watts (1974). White (1948) is a useful source for information on the role of the NAACP in school struggles. Wollenberg (1976) presents an excellent and concise history of racial minorities and schooling in San Francisco. Beasley (1919) is a rich source of information about San Francisco's black community in the late nineteenth century with some material on black schooling. A more recent treatment which makes extensive use of black newspapers and census manuscripts of the period is Lortie (1970). Daniel (1973) is a careful study of the social characteristics of San Francisco's early black community but contains little information specifically relating to schools. The growth of Chicago's black community and the politics of race and schooling is well-documented. Homel (1972) is an exhaustive treatment of the years in which the de facto pattern of segregation emerged. Drake and Cayton (1945) situate the school struggles within the context of the growth of Chicago's black ghetto. Spear (1967) is the best source of events prior to the 1920s.

REFERENCES

BEASLEY, D. L. (1919) The Negro Trail Blazers of California. Los Angeles: Times Mirror Printing and Binding House.

CASTELLS, M. (1978) City, Class and Power. London: Macmillan.
——— (1977) The Urban Question: A Marxist Approach. Cambridge, MA: MIT Press.
COONS, J. (1968) "Chicago," pp. 80-88 in R. Hill and M. Feeley (eds.) Affirmative School Integration. Beverly Hills: Sage.
CUBAN, L. (1976) Urban School Chiefs Under Fire. Chicago: University of Chicago.
DAHL, R. (1957) "Decision-making in a democracy: the Supreme Court as a national policy-maker." Journal of Public Law 6 (Fall).
DANIEL, D. H. (1973) Pioneer Urbanites: A Social and Cultural History of Black San Francisco. San Francisco: R & E Research Associates.
DRAKE, C. S. and H. B. CAYTON (1945) Black Metropolis. New York: Harcourt, Brace, Jovanovich.
ECKE, M. (1972) From Ivy Street to Kennedy Center. Atlanta: Board of Education.
FUNSTON, R. (1975) "The Supreme Court and critical elections." American Political Science Review 69 (September).
HARDIN, R. (1981) Collective Action. Baltimore, MD: John Hopkins University Press.
HARLOE, M. [ed.] (1977) Captive Cities: Studies in the Political Economy of Cities and Regimes. London: John Wiley.
HOMEL, M. W. (1972) "Negroes in the Chicago public schools, 1910-1941." Ph. D. dissertation, University of Chicago.
HORNSBY, A. (1973) "The Negro in Atlanta politics, 1961-1973." Atlanta Historical Bulletin 21.
HUIE, H. M., Sr. (1967) "Factors influencing the desegregation process in the Atlanta school system, 1954-67." Ph. D. dissertation, University of Georgia.
KATZNELSON, I. (1981) City Trenches: Urban Politics and the Patterning of Class in the United States. New York: Random House.
——— (1978) "Considerations on social democracy in the United States." Comparative Politics 11 (October).
——— K. GILLE, and M. WEIR (1981) "Creating public schooling: Chicago and San Francisco, 1830-1870." (unpublished)
KIRP, D. D. (1977) "School desegregation in San Francisco, 1962-76," in A. P. Sindler (ed.) American in the Seventies. Boston: Little, Brown.
LORTIE, F. N. Jr. (1970) "San Francisco's black community, 1870-1890: dilemmas in the struggle for equality." M. A. thesis, San Francisco State.
MOORE, B. (1978) Injustice: The Social Bases of Obedience and Revolt. White Plains, NY: M. E. Sharpe.
ORFIELD, G. (1969) The Reconstruction of Southern Education. New York: John Wiley.
PETERSON, P. E. (1976) School Politics Chicago Style. Chicago: University of Chicago Press.
PICKVANCE, C. G. [ed.] (1976) Urban Sociology: Critical Essays. London: Tavistock.
RABINOWITZ, H. (1978) Race Relations in the Urban South, 1865-1890. New York: Oxford University Press.
RACINE, P. N. (1969) "Atlanta schools: a history of the public school system, 1869-1955." Ph. D. dissertation, Emory University.
Research Atlanta (1973) School Desegregation in Metro Atlanta, 1954-73. Atlanta: Research Atlanta.

SKOCPOL, T. and M. SOMERS (1980) "The use of comparative history in macrosocial inquiry." Comparative Studies in Society and History 22 (April).

SPEAR, A. H. (1967) Black Chicago: The Making of a Negro Ghetto, 1890-1920. Chicago: University of Chicago.

SUBER, M. (1975) "The internal black politics of Atlanta, Georgia, 1944-69." M. A. thesis, Atlanta University.

THOMPSON, E. P. (1978) "The poverty of theory." The Poverty of Theory and Other Essays. London: Merlin Press.

———— (1974) "An open letter to Leszek Kolakowski," in R. Miliband and J. Saville (eds.) The Socialist Register. London: Merlin Press.

TILLY, C. (1975) "Reflections on the history of European statemaking," pp. 3-83 in C. Tilly (ed.) The Formation of National States in Western Europe. Princeton, NJ: Princeton University Press.

VILAR, P. (1973) "Marxist history, a history in the making." New Left Review 80 (July-August).

VON ESCHEN, D., J. KIRK, and M. PINARD (1971) "The organizational substructure of disorderly politics." Social Forces 49 (June).

WATTS, E.J. (1974) "Black political progress in Atlanta, 1868-1895." Journal of Negro History 59.

WEINER, S.S. (1973) "Educational decisions in an organized anarchy." Ph.D. dissertation, Stanford University.

WHITE, W. (1948) A Man Called White: The Autobiography of Walter White. New York: Random House.

WILLIAMS, R. (1961) The Long Revolution. New York: Harper & Row.

WIRT, F. M. (1974) Power in the City. Berkeley: University of California Press.

WOLLENBERG, C. M. (1976) All Deliberate Speed: Segregation and Exclusion in California Schools, 1855-1975. Berkeley: University of California Press.

11

Cultural Identity and Urban Structure:
The Spatial Organization of
San Francisco's Gay Community

MANUEL CASTELLS

KAREN MURPHY

☐ BEAUTIFUL SAN FRANCISCO has become the most promi-
nent urban setting for the expression of the gay culture throughout the
1970s. Gay residents are estimated to be at least 115,000 people; 17%
of the city's population in 1980, two-thirds of them men.[1] They represent a highly visible and mobilized community, able to defend their
identity and to influence the local political system. Although cities
like New York and Los Angeles probably outnumber San Francisco
in the size of the gay population, only in this traditionally tolerant city
have gays been able to build up autonomous social institutions and a
political organization powerful enough to establish a "free commune"
beyond prejudice. Urban researchers and planners should be able to
learn, from such a rich experience, useful lessons about the interaction between city forms and cultural change.[2] Much of San Francisco's housing renovation has come from the gay influx into the city,
along with some of the most vital street celebrations, feasts, and
public events, such as the Castro Fair, the Gay Freedom Parade, or
the new Halloween. On more political grounds, gay supervisors

AUTHOR'S NOTE: The authors acknowledge the invaluable help received for
their research from Richard Schlachman, Bill Krauss, and Harry Britt. We also
thank Lynn Thurston and Richard Lloyd for their research assistance.

Harvey Milk and Harry Britt have been at the forefront of proposals for reform-oriented urban policies, such as rent control, support to low-income housing, and the tax to penalize speculative conversion of residential hotels.

The political strength and cultural freedom of the gay community is, however, actually based upon the spatial concentration of gays in a given city and within certain areas of this city. The voluntary formation of a "gay ghetto" (Levine, 1979) as a space of personal freedom seems to be a precondition for gays to be able "to come out of the closet," a crucial step in the historical process of "gay liberation" (Licata, 1978). The gay *community* can only exist on a territorial basis in which its members might be able to become *visible*, without danger, to express themselves on the basis of a common cultural assumption, overcoming social stigma (Escoffier, 1975). To express their sexual and cultural identity, gays need to occupy symbolically a given territory. In so doing, they deeply transform the city's urban structure. This chapter attempts to analyze the characteristics that allow or resist the process of spatial formation of the gay community, as an exploration of the fundamental relationship between the history of cities and the diversity of human experience.[3]

METHODOLOGY

There is no statistical source that provides information on sexual preferences of residents of specific urban areas, fortunately enough from the moral perspective of the protection of privacy. Yet, such an obstacle appears overwhelming to the researcher trying to understand the spatial dynamics of the emerging gay culture. Thus, our first and main concern has been to establish, as solidly as possible, the precise spatial boundaries of the gay community in San Francisco. Once gayness is related to certain urban units, on the basis of reliable information, it becomes possible to search for potential associations between gay location patterns and different sets of social and spatial variables. In this chapter we will start exploring some relationships between the characteristics of population and housing, and the settlement of gays. It is our hope that when the 1980 census data becomes available, a more thorough statistical and spatial analysis on the reciprocal influence between the evolution of the city and the affirmation of the gay culture will be conducted. Such an analysis will only be possible if it relies on an accurate estimation of the locational pattern of the gay community in San Francisco. To our knowledge such an estimation has never been attempted on any city in the world; it is an additional reason to be particularly cautious in our approach.

Being unable to feel secure enough on any single source of estimation of the gay spatial distribution, *we have used five different sources, obtained in an entirely independent manner.* The fact that all five sources *tend* to show a similar spatial pattern of gay residence and activity reinforces the credibility of each particular source, and provides a very solid support for the mapping that emerges from the five estimations with a very similar profile.

The five sources are the following:

(1) A map of concentration of gay residence established on the basis of *key informants from the gay community.* The main informants were the most qualified political pollsters for gay candidates in San Francisco's local elections. Following our request, informants indicated particular periods of time for gay settlement in each area. Gay residential areas, according to our informants, are cumulative over time. Once they become visible as gay neighborhoods, they do not reverse their character. Actually, at least in San Francisco, they tend to increase their proportion of gay residents after they consolidate as areas of cultural tolerance. On the other hand, it was impossible to obtain any reliable estimation of the actual number of gays in each area. Thus, the areas we characterize as "gay areas" might be so at very different levels, although it is expected that in all these areas there will be a substantial gay residence, and, what is more important, a *visibility of the gay culture.* Yet, it is crucial to keep in mind that we will *not* quantify the presence of gays in particular areas, and that the entire analysis will be based upon the degree of likelihood of each urban unit to be a home for gay men, without considering their numbers.

(2) In a second step of our analysis, we looked for a statistical indicator whose different values could be distributed all over the city. On the basis of direct observation of gay lifestyle, we concluded that an accurate indicator would be *the proportion of multiple male households in each urban unit.* We rejected the proportion of single male households as an indicator, because of the high percentage of nongay elderly living alone. Census data do not provide such information, but the voters' registrar for the City of San Francisco does. We used the 1977 Voters' Registrar data files, the most updated source providing such information. Obviously, the indicator will be the proportion of multiple male households on the total number of *registered voters* in each urban unit. If we were dealing with an ethnic minority population, we would suffer from some uncontrolled statistical bias.

But concerning *gay men,* the source is accurate enough, given their high level of voter registration, in their drive to win electoral power in San Francisco. The assumption, of course, is *not* that all, or the majority, of multiple male households are gay. *The assumption is that there will be a strong correlation between the spatial distribution of gay residence and the spatial distribution of the frequency of multiple male households.* The difference is not merely a semantic one — it will have major consequences for the selection of the statistical techniques suitable to any analysis of this particular data basis.

(3) The third source we selected is *the spatial distribution of the vote for the gay candidate in a city-wide local election.* It is clear that not all people who vote for a gay candidate are gay. It is also clear that there is some close relationship between the number of gays in an area and the vote for the gay candidate in such an area. And, it is equally clear that such a relationship was closer in the early stages of gay mobilization than once gays had already achieved some power on the basis of broader alliances and more diverse constituencies. Thus, we selected as gay strongholds the areas of highest electoral support for the late gay leader, Harvey Milk,[4] in the 1975 supervisorial race, his second attempt at city-wide elected office as a gay candidate and the first in which he obtained a significant number of votes widely distributed across the city (53,000).

(4) The fourth source does not concern residence but *meeting places, particularly bars, health clubs, and entertainment clubs that are aimed primarily at gay patrons.* It has been observed that gay bars tend to overlap with gay residential areas, since they often are the triggers in the settlement patterns in a particular area. To establish the location of gay bars we mapped, according to their street addresses, 232 bars and public places listed by the specialized publication on such matters, the *Bob Damron's Address Book.*[5] We differentiated several points in time, following the publication of the book in different years: 1964-1966, 1969-1971, 1973-1975, and 1980. We then used the list of members of San Francisco's gay "Tavern Guild" Association to check the accuracy of the Bob Damron's book address.

(5) The fifth source concerns the location of *gay businesses.* We mapped 250 businesses (most of them very small) listed in the directory of the Golden Gate Gay Business Association in 1979. Here again, the assumption, relying on our direct observation, is the close connection between gay residence and gay self-proclaimed activities.

To make it possible to establish relationships between these five indicators of "gay territory" and all variables provided by the census, we also transformed the different counting units into census tracts. From now on, these will be our units of observation and analysis.

1950s.

Early 1970s.

1960s.

Late 1970s.

Mid 1970s.

Map 11.1   Gay Residential and Social Areas: Sequential Development of Gay Residential Areas, San Francisco, 1950-1980

SOURCE: Key informants from the gay community.

NOTE: In 1980 *all* marked areas are characterized by significant gay residence.

On the basis of these five *different independent sources,* we established five maps, as presented in the following section. The gay concentration areas that result from the observation of these five different maps tend to fit very closely into a common spatial location within the city of San Francisco.

To rely on a somewhat less intuitive measure, we calculated the zero-order correlation between the spatial distributions of the indicators of gay *residence* we proposed. To do so, we gave dichotomic values (1,0) to areas with gay presence or absence (Map 11.1); we proceeded in a similar way classifying in 1 versus 0 the areas with high

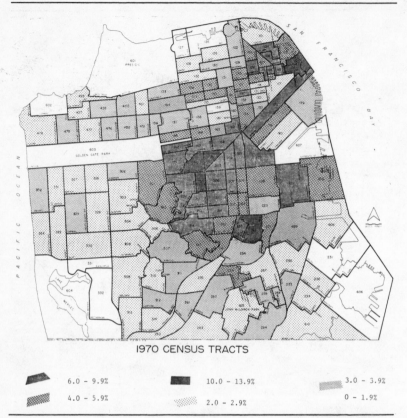

1970 CENSUS TRACTS

	6.0 – 9.9%		10.0 – 13.9%		3.0 – 3.9%
	4.0 – 5.9%		2.0 – 2.9%		0 – 1.9%

Map 11.2   Gay Residential Areas, as Indicated by the Proportion of Multiple Male Households Over the Total Registered Voters in Each Census Tract, 1977

SOURCE: 1977 Voter Registration Tape/Computer Program prepared by Doug De Young.

NOTE: Voter registration household median = 3%; 1970 census nonrelative households citywide = 3.6%.

or low proportion of multiple male households (levels 6, 5, 4, versus levels 3, 2, 1, in the scale of census tracts in the city according to the proportion of multiple male households, as shown in Map 11.2). We then calculated the correlation for all census tracts distributed in relationship to the values (1,0) of the two indicators. We proceeded in the same manner with the third indicator, namely, the importance of gay vote.

The correlations are highly positive: $r = .55$ ($p < .001$) between the areas designated by the community informants and the ones resulting

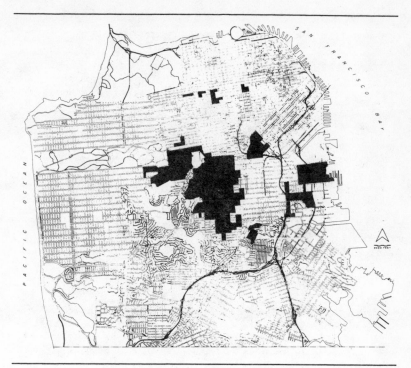

**Map 11.3   Gay Voting Pattern: Two Top-Ranking Precincts in Support of Harvey
Milk (Gay Candidate) in the 1975 Supervisorial Race**

SOURCE: Harvey Milk Campaign Office.

from the spatial distribution of multiple male households. And $r = .67$
($p < .001$) between the definition proposed by the informants and the
distribution of the 1975 vote for Harvey Milk. Statistical measures
confirm what the simple observation of the maps suggests: We have a
similar pattern of spatial location provided by different independent
indicators whose credibility becomes, thus, mutually reinforced. We
now are able to define the spatial boundaries of San Francisco's gay
male community *as of 1980*. Also, given the verification of the accuracy of the estimation provided by our key informants, we may, unless
contrary evidence is provided, consider as a probable trend the spatial
sequence described by them over time.

Having established the spatial profile of the gay community, we
are now able to relate it to social and urban characteristics distributed
across the city, in order to understand the factors fostering or coun-

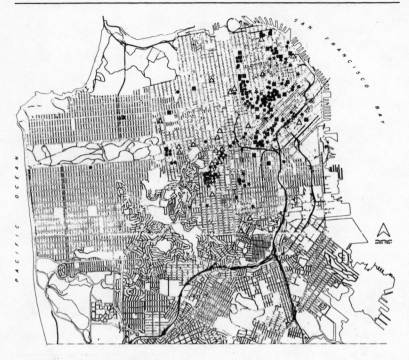

■ 1964-66 (58 establishments).

★ 1969-71 (73 establishments).

● 1973-75 (147 establishments).

△ 1980 (232 establishments).

**Map 11.4 Gay Gathering Places (Including Bars and Social Clubs), Sequential
Cumulative Development**

SOURCE: Bob Damron's Address Book. (Map constructed on the basis of a series of
maps by Michael Kennedy.)

teracting the patterns of settlement inspired by the gay culture. On the
basis of the 1970 U.S. Census and of the 1974 U.S. Bureau of the
Census *Urban Atlas*, we have selected eleven variables considered to
be relevant to our analysis, as presented in Table 11.1 (U.S. Bureau of
the Census, 1970, 1974).

Yet we must keep in mind that the only thing this particular data
base tells us about gayness is *where* gays are or tend to be. We do not
have any indicator of *gayness*; the only thing we have is a series of
converging indicators of gay *location*. Thus, we cannot infer anything
about gay *individuals*. We only can analyze gay versus nongay *spatial*

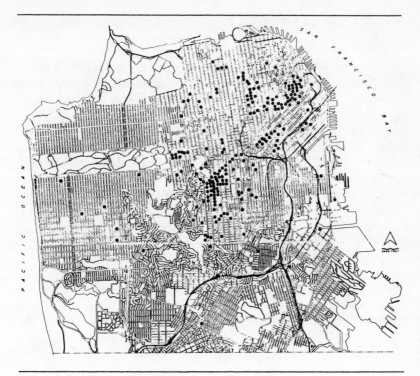

Map 11.5   1979 Golden Gate Business Association Members
SOURCE: GGBA Buyers' Guide/Directory, Fall-Winter 1979-1980.

units. To proceed in such a way we have carried on two different analyses:

(1) On the basis of our *key informants map,* we have divided the city in two categories of spatial areas: those with gay presence, and those without such a presence. We have calculated for each selected variable the mean value of their distribution in the census tracts corresponding to each one of the two categories of space. The T test provides the statistical significance of the differences observed for the mean value of each variable in each one of the two categories of spatial units. The extent and the direction of such differences will provide the clue for the social and urban specificity of the gay territory.

(2) We have tried to establish some measure of correlation between the spatial distribution of the gay community and the spatial distribution of selected social and urban variables. Since we are *not*

looking for covariation between characteristics across the space, but for correspondence between spatial organization in relationship to two series of criteria, the usual regression analysis is inadequate to our purpose. Instead, we have proceeded to calculate rank-correlation coefficients (Spearman test) between, on the one hand, the distribution of all census tracts in a 6-level scale of "gay space," constructed on the basis of the proportion of *multiple male households;* and, on the other hand, the distribution of all census tracts in a series of 6-level scales on the basis of selected variables. Only variables that the first step of the analysis (as described above) showed to be discriminatory, were considered to construct the scales of social and urban differentiation in San Francisco. Furthermore, one of the findings of our analysis is that these selected variables are likely to affect gay location patterns *jointly,* instead of having individual effects, since their spatial distribution does *not* entirely overlap. Thus, to test our analysis we constructed an additional scale, classifying the census-tract units in a scale that combines in its criteria the three variables considered to have some effect on gay location. The Spearman coefficient between the ranks obtained by the census tract units in this scale and the ranks they obtained in the multiple male households scale, provides the final test for our tentative interpretation of the social roots of gay location patterns.

THE FINDINGS: PROPERTY AND FAMILY
VERSUS THE GAY CITY

The most important result of our study is the definition of the spatial organization of the gay community as presented in Maps 11.1, 11.2, 11.3, 11.4, and 11.5. All converge toward a largely similar territorial boundary. Thus, the first finding is that *there is a gay territory.* Furthermore, it is not only a residential space — it is also the space for social interaction, for business activities of all kinds, for leisure and pleasure, for feasts and politics. It is on such a spatial basis that, using the decentralized electoral system by districts that existed in San Francisco between 1976 and 1980, the gay community succeeded in electing both Harvey Milk and Harry Britt, to become the largest and best-organized single voting block. Such a connection between an autonomous culture, small business, local political power, along with the open expression of sexuality on the basis of a space of freedom opposed to the dominant social rules, offers a striking parallel with the emergence of urban culture in the European free cities of the late Middle Ages (Weber, 1958). The "gay city" appears as a *contiguous urban space* with its own norms and institutions, a city within the city.

What are the characteristics of such a specific space? Table 11.1 provides a tentative answer, according to the methodology we previously presented. Of the eleven selected variables examined, only two appear *strongly negatively associated* with the presence of gays: the *proportion of owner-occupied housing units,* and the *proportion of the resident population under 18 years of age*. In other words, *property* and *family* are the major walls protecting the "straight universe" against the gay influence. It is our hypothesis that what appears fundamental in this relationship is the cultural dimension of the rejection and the capacity of a given neighborhood to oppose gay immigration. It can do so either because of control over real estate property or because of the local organization of a family-oriented community, as is the case in the Latino Inner Mission District and in the black Bayview-Hunters Point area, which basically kept outside gays' location patterns. Variables in housing value, family income, or blue-collar occupation do *not* differentiate gay and nongay areas: neither gay location nor gay rejection appear to be strictly economic processes. They come basically from cultural distances. Two other variables show some significant difference between the city and the gay city: Gay areas tend to concentrate a more educated population; and gay residence is less frequent in high rent areas. One should remember that San Francisco's housing stock is 65 percent rental and that the highest social status occurs in areas like Pacific Heights, where the proportion of owner-occupied housing is below the average. Thus, in areas exclusive enough to keep gays out by the means of high rents, a new limit is established.

These findings are reinforced by the second method we have used to measure our observations. Table 11.2 shows that, as expected, the rank-correlation between the spatial distributions of our indicator of gay residence and each one of those for the four variables obtained by our previous analysis are almost nonexistent. A simple observation of Map 11.6 will provide the key. Gays are not present in the western slope of the city because of the influence of property; in the eastern side because of the influence of family; and in the northern hills because of high rents. Only when we combined the effects of the three variables in one scale (*property* and *family* and *high rents*), did we obtain a significant -.38 rank-correlation coefficient between this scale and the scale constructed on the basis of the indicator of gay residence.

To provide a more clear expression of these results we have constructed Map 11.6 in which the three variables and the proportion of multiple male households are mapped *together* according to the description in the map's legend. Map 11.6 shows a striking autonomy

TABLE 11.1 Statistical Difference in the Values of Selected Variables for Spatial Areas with Significant Gay Residence and without Significant Gay Residence, San Francisco, circa 1970.

1	2	3	4	5	6	7
				Pooled Variance Estimate[a]		
Variable	Cultural-Spatial Area	Mean	Standard Deviation	T Value	Degrees of Freedom	2-Tail Probability
Family						
% of the resident population under 18	Non-Gay	22.8	10.56	4.25	146	.000
	Gay	15.9	8.43			
Property						
% of housing units owner occupied	Non-Gay	40.54	31.51	4.41	146	.000
	Gay	20.93	18.02			
Education						
% of high school graduates among residents aged 25 and over	Non-Gay	57.21	20.05	−1.55	145	.123
	Gay	61.73	13.08			
Housing Rents						
Median Contract rent, 1970	Non-Gay	135.04	44.36	1.37	144	.174
	Gay	126.16	29.76			
Housing Value						
Median housing value, dollars, 1970	Non-Gay	26,886	12,003	−.26	122.5	.795
	Gay	27,384	9,360			
Income						
Median family income, dollars, 1970	Non-Gay	9,858	3,989	.82	144	.415
	Gay	9,322	3,804			

				t	n	p
Blue Collar % of blue collar employed among residents	Non-Gay	22.73	12.44			
	Gay	21.72	9.32	.54	146	.588
Black Population Blacks as percent of total population	Non-Gay	16.17	25.98			
	Gay	11.96	17.98	1.10	146	.273
Elderly % of total population over 65.	Non-gay	14.66	9.51			
	Gay	14.45	6.29	.15	146	.879
Housing Dilapidation % of housing units lacking some or all plumbing	Non-gay	7.68	18.09			
	Gay	8.75	13.81	− .39	145	.695
Housing Decay % of housing units lacking kitchen facilities	Non-gay	7.66	18.45			
	Gay	10.89	17.60	−1.07	145	.288

SOURCES: U.S. Census, 1970, San Francisco. Gay and nongay areas have been defined by classifying in two separate categories all census tracts according to information provided by key informants of the gay community, and checked with four other sources, as described in the chapter.

[a]Basic criterion for the statistical significance of the observed differences between mean values of each variable has been the 2-tail probability threshold as provided by the T test. P < .100 has been considered significant; p values between < .1000 and < .2000 have been considered moderately significant; p values > 2000 have been considered randomly produced, although they *do exist as trends* of the urban reality of San Francisco.

TABLE 11.2 Spearman Correlation Coefficients Between the Rank of Census
Tracts in San Francisco Ordered on a Six-Level Scale According
to the *Proportion of Multiple Male Households Among Registered
Voter Residents*, in 1977, and Five Other Scales as Indicated

	Spearman R
1. Scale according to the proportion of residents under 18 in 1970 (FAMILY) (6 levels)	−.11
2. Scale according to the proportion of owner-occupied housing units in 1970 (PROPERTY) (6 levels)	−.16
3. Scale according to the proportion of high-school graduates among residents over 25, in 1970 (EDUCATION) (6 levels)[a]	−.18
4. Scale according to the median contract rent, in dollars, in 1970 (6 levels)	−.17
5. Scale of FAMILY, PROPERTY and HIGH RENT. (Scale in 6 level constructed by combination of scores of census tracts in scales number 1, 2, and 3. It reads as follows: top level in scale 5 = top level in scale 1 *or* in scale 2 *or* in scale 3, and so on throughout all the levels of the scale.)	−.38 (p < .001)

[a]The ranking for *education* has been inversed in order in relationship to the other scales, to
be consistent with the hypothesis of a *direct* correlation with gay presence.

between the territories defined by the four variables, with only some
very limited overlapping. We are in the presence of four spaces:
property land, family land, rich land (defined by the proportion of
high-rent housing), and gay land. Gay settlement is opposed by
property, family, and the high class: the old triad of social conservatism.[6] Map 11.7 gives a good confirmation of such a conclusion.
Our informants (whose accuracy was established through cross-
checking with the four remaining sources) forecasted the expansion of
the gay community throughout the 1980s in the areas presented in
Map 11.7, without any previous knowledge of our analysis. They are
precisely the zones that appear to be contiguous to the gay territory
and where the three variables proposed as deterrents to gay location
score relatively low values. The gay city seems to evolve according to
a social logic whose historical sequence we will now try to explain.

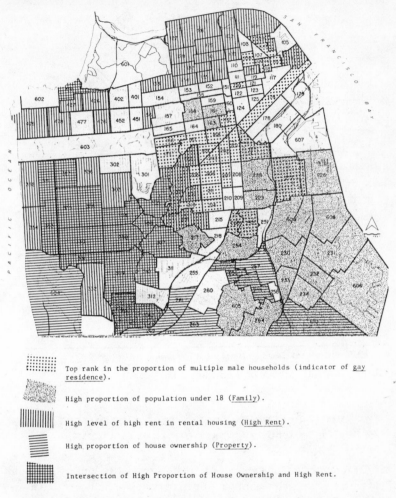

Top rank in the proportion of multiple male households (indicator of gay residence).

High proportion of population under 18 (Family).

High level of high rent in rental housing (High Rent).

High proportion of house ownership (Property).

Intersection of High Proportion of House Ownership and High Rent.

SOURCE: Voters' Registrar, 1977; U.S. Census *Urban Atlas* (1970 Data), San Francisco, 1974.

NOTE: Characterized by high scores on 4 different scales: (1) Proportion of multiple male households, 1977; (2) home ownership, 1970; (3) proportion of residents under 18, 1970; (4) high level of median contract rent, 1970. (Census tracts can overlap since the definition of their rank in each one of the 4 scales is theoretically independent.)

Map 11.6   Family, Property, High Rents, Gay Residential Areas

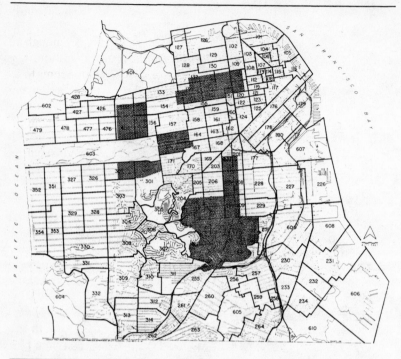

Map 11.7   Informant Basemap Residential Expansion Areas (1980s)
SOURCE: Key Informants Mapping (February 1980).

WHY SAN FRANCISCO? THE HISTORICAL CONSTRUCTION OF THE GAY CITY[7]

An instant city, settlement of adventurers attracted by the gold fields, San Francisco was always open to personal fantasies and loose moral standards (Becker, 1971). A port city, its Barbary Coast was a meeting point for sailors, travelers, transients, and lonely persons, a milieu of casual encounters and low social control, where homosexuality could express itself because of the blurred borderline between the normal and the abnormal. A free city, a gateway city, on the western limit of the western world. Only there, on the marginal quarters of a marginal city, homosexuality could blossom. And yet, even there, "deviants" were repressed and forced into hiding, when San Francisco decided to become respectable, to emerge as the moral and cultural capital of the West and to grow up gracefully. The reform movement finally reached the police in the 1930s and forced a parallel crackdown on prostitution and homosexuality, the (significantly enough) twin evils of puritan morality (Wirt, 1974).

Thus, the pioneer origins of San Francisco are not enough to explain its manifest destiny as the spatial setting of gay liberation. The major breaking point in this particular historical evolution seems to be World War II. San Francisco was the main port of embarkation and disembarkation for the Pacific front. An estimated 1.6 million men and women went through the city — young, alone, suddenly uprooted, living on the edge of death and suffering, living together with people of the same gender. The average 10 percent of homosexuals in all human populations (Kinsey, 1948, 1958) discovered themselves more easily and more rapidly in such a context. Others discovered also the potentials of bisexuality. Once again, Thanatos and Eros played into each other. Many servicemen and women were punished in the Army and the Navy, over 40,000 known cases. Most of those dishonorably discharged from military service in the Pacific were disembarked in San Francisco. No way to return home bringing with them the stigma. They stayed in the city. They were reinforced at the end of the war by many others that, having secretly discovered their sexual and cultural identity, did not want to live any more in the small towns of Middle America. They met in bars, particularly in the Tenderloin, the nightlife zone adjacent to the downtown area (see Map 11.4). Bars were then, before, and after, the focal points of social life for gay people. Networks were constructed around bars, and some specific forms of culture and ideology emerged occasionally in such places (Weightman, 1980).

So gays were forced to be a part of the relatively tolerated milieu of individual "deviant" behavior. There was no community, but networks. No territory, but places. In the 1950s, an alternative lifestyle culture, the *beatniks*, flourished in the cafes and bookstores of North Beach (see Map 11.1). Gays found there a receptive, tolerant atmosphere. They were a part of the "rebel" residential community that established itself in that area. Without becoming a community by themselves, they enjoyed some relative freedom as a subset of a highly intellectual counterculture. In the 1960s, gays' networks expanded and took advantage of the atmosphere of liberalism imposed by the antiwar and civil rights movements. Bars flourished around a new axis, Polk/Van Ness, more integrated into the city, in addition to the two areas already marked by the counterculture (Tenderloin and North Beach — see Map 11.4). For the first time, a gay residence developed on its own around the meeting places opened in Polk Street. When the adjacent ghetto of the Western Addition started to be disintegrated by the Urban Renewal Program, some gay households started buying property in the renewed area in the late 1960s (see Map 11.1; see also Butler, 1978).

Afterwards, the Gay Liberation movement took a dramatic impulse with New York's "Stonewall" revolt of June, 1969 (Humphrey, 1972). Gays came out of the closet and many of them, all over the country, migrated to the new Meccas where they could express their identities: New York, Los Angeles, San Francisco. This time it was not only a matter of being close to social places where sexual networks could be connected. It was a whole specific lifestyle, *defined as gay,* that the new *militants* tried to develop symbolically and politically. They tried to establish a *community,* in its full multidimensional meaning. Others joined the effort in a somewhat confused way, to learn about themselves, to be *socialized* within a new culture more suitable to their needs and behavior, a culture that they needed to feel by being there, in a given neighborhood, as they learned what they were *not* by being in the classrooms of their hometown's schools. So, a collective movement, informally organized, started to take over a very specific area (the Castro Valley, in the geographical center of San Francisco), an area characterized by two main features: It was *not* a traditionally "deviant" area, but a beautiful old-Victorian neighborhood partly vacated by its Irish working-class dwellers moving to the suburbs; it was a very middle-level area in terms of income and education but there was a large proportion of affordable *rental* housing (Lee, 1980). Gays purposefully started businesses and stores in the area, and organized a neighborhood merchant association that forced other businesses in the area, particularly banks, to collaborate with them. Bars and social places followed the residential concentration, reversing the trend: *Community substituted for networks.*

At the same time, in the early 1970s, two very different trends were expressing other different spatial and social orientations of gay people, trends less well known because of the too-bright lights of the Castro's self-constructed ghetto. On the one hand, many gays, while sharing the aspiration of a self-controlled "private territory," could not afford the high rents that realtors were asking gays to pay to have some feeling of freedom (Lee, 1980). Thus, they started another "colonization" in the much harsher area of South of Market, where transient hotels, warehouses, and slums waited for redevelopment. Their marginality from the gay community was not only spatial. Socially, they tended also to reject the politicization and positive counterculture of the new liberation movement. They emphasized the sexual aspects of the gay condition. The more the gay community appeared in a process of legitimation, the more a strongly individualistic minority, generally poorer and less educated, headed toward self-affirmation of a new sexual "deviance," many of them joining the sado-masochist networks: South-of-Market became the quarters of the "leather culture" (Dreuilhe, 1979).

On the other hand, a growing proportion of middle-class gays rejected the militant stands of the liberation movement and considered ridiculous the idea of setting up a ghetto. Their way was to obtain personal freedom and legal rights without challenging a system that otherwise was highly favorable to their sex, to their class, and to their race. Such an orientation had also a distinct spatial expression. According to our informants, in the early 1970s, middle-class gays, wanting to be in a friendly gay neighborhood but rejecting the Castro ghetto, started to locate in the lower sections of fashionable Pacific Heights, literally on the threshold of the heaven of San Francisco's elite, in a highly symbolic move.

In the mid-1970s, two major spatial developments characterized the gay community: (1) The Castro ghetto grew and expanded dramatically in all adjacent areas. A very dense network of gay bars, health clubs, stores, business, and activities developed on the basis of such a growing market. Immigration from the rest of the country accelerated. Gay households started buying property in this neighborhood, taking advantage of the process of conversion of rented buildings into condominiums. They moved westward to more affluent sections. The expansion also reached the Dolores Corridor on the border with the Latino Mission District. Frictions appeared and some antigay violence surged among the Latino youth. (2) In a different development, gay households (particularly lesbian) settled in the working-class areas of Bernal Heights and Potrero Hills, within what we have called *family land*, taking advantage of the existence of relatively sound cheap housing. The major reason for such a tolerance was the existence in those areas of very large countercultural communities, established there by young white students and professionals who maintained a good relationship with working-class and minority families on the basis of their militancy in favor of the neighborhood. Gay men and lesbians were generally welcomed by such a preexisting milieu of 1960s rebels.

Finally, in the late 1970s, on the basis of the newly acquired relative protection through political power, the new expansion of gay residences took place in the borders of middle-class areas, where home ownership was generally required: Inner Richmond and Inner Sunset neighborhoods in the western part of the city. While still preserving its own territory, gay residences started to overcome the last barriers to their spatial presence. The contradiction implicit in such a move is that either gays will lose their identity to adapt to prevailing patterns of behavior, or they will have to obtain from the entire city a degree of tolerance that largely exceeds the capacity of a given society to challenge the fundamental values of family and sexual repression.

CONCLUSION: THE SOCIAL FUNCTIONS OF THE GAY TERRITORY

The gay community expresses a diversity of interests and social situations. It is at the same time *a sexual orientation, a cultural lifestyle,* and *a political "party."* Besides, in the case of San Francisco, a very important sector of the gay people is formed by *young professionals and small businessmen* emigrated to San Francisco both because of its cultural tolerance and because of its character as a headquarters of an advanced service economy. In fact, all these aspects are present, with different weights in each period, in the spatial history of the gay community.

When the gay *identity* had to be reduced at its hard core, because of fierce social repression, sexually oriented networks were the basic instrument of communication and solidarity. But gayness is more than a sexual orientation: It is an element of a broader culture aimed at posing an alternative way of life, characterized by the domination of expressiveness over instrumentalism, and by human interaction over impersonal competition. Such values were very close to the beatnik and hippy cultures of the 1950s and 1960s. This is why gays became incorporated into the territories of the alternative lifestyles. But they were different and they were more than a counterculture. Not only did they have a specific sexual network to preserve. They had to fight political battles, they had to change laws, to resist police harrass-ment, to influence government. The problematic therefore arises as to how to organize within existing political institutions principally on the basis of bars and marginal countercultures. To be a society within the society, they had to spatialize their oppression, to transform it into an organizational basis to achieve political power. This is why the build-ing of the Castro ghetto is inseparable from the development of the gay community as a social movement: It brought together a sexual identity, a cultural self-definition, and a political project, on the basis of a specific organizational form relying on the control of a given territory. This is why those who rejected the building of a social movement quit the "commune," either to rejoin the mainstream society in exchange for some tolerance, or to refuse all communication in the name of irreducible affirmation of individual sexual pleasure, in an even more destructive escape toward the sterile lands of self-generated excitement. The ghetto preceded the moment of institutionalization. It provided the basis, particularly through decentralized district elections, to obtain enough power to live in the city instead of just being safe within the community. Yet, as long as gays are insulted, beaten, and killed because of whom they love, even in the city where they have access to institutionalized power, they still need the ghetto (namely, the territorial boundaries of a cultural community) for the same reasons that Jewish people in Europe, black people in America, and oppressed ethnic minorities all over the world, always needed it — for everyday's survival.

NOTES

1. Estimations of the San Francisco gay population (obviously unrecorded in the Census data) vary according to different sources. Deborah Wolf, in her book *The Lesbian Community* (University of California Press, 1979) affirms that "it has been estimated that by 1977 about 200,000 homosexual women and men . . . live in San Francisco at any one time" (p. 74). Claude Fischer, professor of urban sociology at the University of California at Berkeley, on the basis of several sources including his own survey, considers that "a more accurate estimate is that perhaps 12 to 15 percent of all voting-age population *adults* in the city are homosexual. Nevertheless, the visibility and power of homosexuals exceeds this deflated figure. They probably represent 20 percent or more of the white, anglo population, and may comprise 25 to 30 percent of all white anglo *men* living in the city" (Fischer, 1981: ch. 2). Most political analysts, including gay voting specialist Richard Schlachman, believe gays represent 25 percent of registered voters, that is about 95,000 individuals. Yet, while gays tend to be very active in the political system, a large proportion of the city's ethnic minorities (including noncitizens) do not register as voters, and yet there is *some* gay presence among them that should slightly increase the 95,000 figure. In fact, the only serious way of obtaining a reliable estimation would be to ask a representative sample of the population their sexual preference. Several campaign polls conducted both statewide and locally have questioned voters as to their sexual preference. The most recent San Francisco survey (unpublished) having included such a question in the poll was conducted by Professor Richard De Leon, of the San Francisco State University's political science department, for the City Charter Commission of the City of San Francisco. On a representative sample totalling 1,337 individuals, the findings seem to indicate that the total gay population would be 17 percent. For gay men, it would be *much higher*: 24.4 percent. To stay at the level of the global population, where more estimates are done, 17 percent of the estimated 678,940 population in 1980, representing 115,419 individuals, corresponds, very roughly, to the commonly cited figure of *about 100,000*, and to a somewhat higher figure of the estimation of registered voters. Therefore, we can reasonably consider

that *the gay population in San Francisco must be between 100,000 and 120,000 individuals, (two thirds of them being gay men and one third lesbians)*, according to De Leon's survey.

2. In the well-established intellectual tradition so vigorously defended by Lewis Mumford.

3. The lesbian community has not been considered in this analysis, in spite of some accurate information we obtained on its spatial organization. The reasons are twofold: (1) The lesbian territory follows a different logic; (2) the control over a given space is much less important in the process of lesbian liberation, more oriented toward radical cultural change than toward the creation of a basis of power within existing institutions. For a confirmation of this point of view, see Wolf, 1979.

4. Murdered in 1978, jointly with Mayor Moscone, by a radical right-wing fellow supervisor,

5. *Bob Damron's Address Book*, now in its seventeenth year of publication, has long been considered one of the most comprehensive and reliable national guides to gay bars, baths, discos, hotels and restaurants. With over 5000 listings in the 1981 publication, the *Address Book* provides the user with both the location and a brief description of the establishment in a symbolized format. The accuracy of the San Francisco listings can be considered greatly enhanced, given the location of the central office in San Francisco.

6. The indicator of "high rent" per se does not express "high class". Yet, an observation of Map 6 clearly reveals that, *in the case of San Francisco,* very high rents in a predominantly rental housing market are only an effective deterrent against gays in the upper-class section of the city, Pacific Heights. While the level of income does not reveal this tendency, the high rents, outside the homeowner-occupied areas, do.

7. The historical analysis established by the authors through in-depth interviewing of historical witnesses. Very insightful analysis and information was provided by Escoffier and D'Emilio, 1981.

REFERENCES

BECKER, H. [ed.] (1971) Culture and Civility in San Francisco. New Brunswick, NJ: Transaction.

BUTLER, K. (1978) "Gays who invested in black areas." San Francisco Chronicle (September 1).

DREUILHE, A.E. (1979) La Societe Invertie ou les Gais de San Francisco. Ottawa: Flammarion.

ESCOFFIER, J. (1975) "Stigmas, work environment and economic discrimination against homosexuals." Homosexual Counseling Journal 2 (January).

———— and J. D'EMILIO (1981) "San Francisco's gay history." Socialist Review.

FISHER, C.S. (1981) To Dwell Among Friends. Chicago: University of Chicago Press.

HUMPHREYS, L. (1972) Out of the Closets: The Sociology of Homosexual Liberation. Englewood Cliffs, NJ: Prentice-Hall.

KINSEY, A. et al. (1958) Sexual Behavior in the Human Female. Philadelphia: W.B. Saunders.

———— (1948) Sexual Behavior in the Human Male. Philadelphia: W.B. Saunders.

LEE, D. (1980) "The gay community and improvements in the quality of life in San Francisco." MCP thesis, University of California, Berkeley.

LEVINE, M. (1979) "The gay ghetto," in Martin Levine (ed.) Gay Men: The Sociology of Male Homosexuality. New York: Harper & Row.

LICATA, S. (1978) "Gay power: a history of the American gay movement, 1908-1974," Ph.D. thesis, University of Southern California, Los Angeles.

U.S. Bureau of the Census (1974) Urban Atlas. Washington, DC: Government Printing Office.

——— (1970) "Census of population and housing, 1970." PHC (I) Series, Census Tracts. San Francisco-Oakland, California.

WEBER, M. (1958) The City. Glencoe, IL: Free Press.

WEIGHTMAN, G. (1980) "Gay bars as private places." Landscape 24, 1.

WIRT, F. M. (1974) Power in the City. Decision-Making in San Francisco. Berkeley: University of California Press.

WOLF, D. C. (1979) The Lesbian Community. Berkeley: University of California Press.

12

Politics, Parties, and Urban Movements: Western Europe

PAOLO CECCARELLI

☐ IN THE LATE 1960s and early 1970s, large European cities experienced an unprecedented outburst of urban movements. The Spring of 1968 in Paris and in West German cities, the "Hot Autumn" of 1969 and the large-scale urban struggles of the early 1970s in Italy; squatters' movements in Portuguese cities in the aftermath of the April Revolution; urban social movements in Spain at the end of Francoism — all are examples of this massive grassroots political mobilization which sprang forth very rapidly, and almost unexpectedly, all over Europe. Although they occurred only a short while ago in the context of the present political situation, these enormous events seem so remote and peculiar that one might wonder whether they actually took place.

Large-scale urban social movements have faded out as rapidly as they originated and no longer play the role that they did in the political process of the past decade. This is so in spite of the fact that most of the problems which at the end of the 1960s had supposedly ignited urban movements are still unsolved or even aggravated. Unemployment has increased; the cost of living has skyrocketed; social services have worsened because of cuts in public expenditure; there is a shortage of adequate housing and what exists is very expensive. Another target of social movements of the 1960s and 1970s — political authoritarianism — is still strong and alive. Thatcherism in the United Kingdom, the failure of left-wing coalitions in Portugal, the resurgence of conservative parties and the military in Spain, the inde-

structibility of "Centrismo" in Italy suggest that the political scene
has not changed much since the Paris May. What was once hailed as a
social, political, and cultural turning point in the post-World War II
Europe is at present frequently under criticism. It is often interpreted
as a temporary break in a steady process of political evolution — a
juncture at which a real quest for change combined with foggy
idealism and obsolete left heresies; an unsuccessful attempt by a
frustrated middle-class intelligentsia which had been put aside in the
historical process to regain political legitimacy.

What does the present eclipse of urban movements mean? Why
and how are they at the present standstill? It is simplistic to believe
that social and economic conservatism, fear of economic crisis, rigid
controls over public expenditure, and the ability of the state to contain
social struggles both by positive and repressive action, have put an
end to the process of innovative political mobilization that an insurgent decade had started.

Why is the political and ideological relevance of urban movements
so often dismissed? Why are the traditional party system, the established structures of representative democracy, and even neocorporatism back in fashion again? Even if urban social movements are
more limited in size and strength, fragmented, and have a lesser
impact now than they had in the political process of past years, they
have not disappeared. In fact, political mobilization has changed in
scope, and centered on more general and basic issues, such as the
anti-nuclear movements in West Germany and France, and the coalitions to support abortion in Italy. And while it has declined in some
countries, political mobilization has emerged in new forms elsewhere.
This is the case of the outburst of urban conflict in Switzerland and the
reemergence of housing struggles in the Netherlands and more recently in West Berlin. At least latent forms of grassroots political
mobilization, still radical in nature, continue to exist.

However, the continuous reemergence of urban movements, no
matter how marginal and fragmented, does not seem to imply a kind of
"backyard revolution" European-style (Boyte, 1980). They cannot be
explained as a continuous flow of circumscribed political actions with
occasional upsurges, with roots in participation in local government,
in reformist politics, in a long fight for self-government to better
protect local interests against a centralized and authoritarian state.
Urban movements have frequently turned into mass political mobilization centered on broader issues. And recent scattered urban
movements are often more radical and violent in nature than the ones
of the 1960s and 1970s. How does this political mobilization survive?
And how does it relate to political parties and their strategies?

In this chapter, I shall try to answer some of these questions. I argue that it is impossible to work out a comprehensive explanation of the present decline of urban movements in European cities. Recent events make us aware of the fact that "crystal ball" forecasts about processes which have only had a short span of time to develop are risky. Urban social conflicts might suddenly start again, eluding our expectations and theories, as happened at the end of the 1960s. Even when urban movements declined rapidly in the mid-1970s, they did so denying the theories that authoritative social scientists had worked out.

NATURE OF URBAN SOCIAL MOVEMENTS

According to a number of current assumptions, urban social movements are one of the several indicators of the impending crisis of traditional systems of consensus-making and representation in Western bourgeois democracies. When they began at the end of the 1960s, they anticipated, and then contributed to, the acceleration of this crisis (Castells et al., 1974; Della Pergola, 1975; Borja, 1975). Conflicts spread everywhere, centering on crucial aspects of late capitalist societies — the role of the state and of political elites, reproduction and accumulation processes, the condition of women in society, and so on. It appeared to be almost impossible to control them. In addition, these conflicts ran parallel to a massive wave of similar labor disputes. Several slogans of labor struggle were shared by actors in the urban movements — egalitarianism, self-government, and an end to the division of labor. It was assumed that should urban and labor conflicts combine together, their impact would produce a major crisis for western capitalist societies and economies.

A number of facts seemed to confirm this assumption. Two among them are worth mentioning:

(1) At the end of the 1960s, Western European left-wing parties faced increasing difficulties in their efforts to adjust to the new needs and attitudes of the changing European working class. Old paradigms, traditional ideologies, strategies and organizational assumptions on which the struggles against fascism and the post-war capitalist reconstruction had rested were no longer adequate. Similar problems were faced by the unions. The only viable solution seemed to be a substantial renovation of left-wing parties, and of their basic politics.[1]

(2) European economies faced increasing problems, and a number of them were already experiencing serious structural difficul-

ties. Growth rates sustained during the 1960s slowed down dramatically in the early 1970s. The economic system appeared to be
very rigid and unable to face change. The economic policies of the
state failed to cope with the ever-increasing public expenditure on one
side and the steady decline of productivity and competitiveness in
international markets on the other. Entrepreneurs pressed for a
thorough industrial reorganization at the expense of labor, but this
clashed with the strategy of the unions and the new militancy of the
workers. Apparently, there was little or no room for maneuvering,
and chances for reformist economic policies were nil. A substantial
change in economic policies and labor relations appeared to be unavoidable.

I do not consider these pieces of evidence at all adequate. The
assumption that urban social movements anticipated the process of
change of the 1970s is grounded on a set of interpretations of the
existing social and economic situation that are too general to be fully
convincing.

Other explanations ought to be introduced, among them the
hypothesis that the process of political and social change in European
societies was already on its way when the urban social movement
exploded. It is possible that urban social movements and the conflicts
which accompanied them, far from anticipating a new era, were the
expression of the last and most conflictive stage of a process of change
and readjustment to it. Mass political consciousness was eventually
aroused because changes were scattered, insufficient, and inconsistent, and in fact exposed and magnified social inequalities and dysfunctions. At that point, the political and social systems were ready
for a change and had begun the process of readjusting to the new
needs. This does not mean that these systems would necessarily
undergo such transformations, even lacking mass political mobilization and confrontation. It does mean that urban social movements do
not by definition always anticipate and ignite change. Such an
hypothesis might better explain why urban social movements faded
out rapidly, and why a new social pact between power elites, left-wing
parties, and the unions was rapidly agreed upon at the end of the
1970s. In other words, it is possible that at the end of the 1960s, the
economy was already under reorganization in order to survive a
period of severe recession and highly uneven performances in different sectors of the economy.

The state also was in a process of restructuring. And the decentralization both of decision- and policy-making could result in the
anticipation and eventually the stifling of conflicts. Society was un

dergoing change as well. It had broken into a variety of subsystems, the features and the roles of which were increasingly difficult to classify into traditional typologies. Of course, this is not the place in which to develop this hypothesis in all its theoretical and factual implications. In this chapter, I intend to provide a few elements for further thinking and debate.

I shall first examine briefly some aspects of the interrelation between urban movements and political parties, and between urban movements and structural transformations in the economy. I shall subsequently analyze in more detail one "case study," focusing on Italy, where these relationships appear to have taken an extreme form. I want to point to a set of facts which allow us to better understand the difference between the present social and political context and the context in which those movements flourished a decade ago. This will also help to cast a different light on the nature of such movements and possibly suggest alternative explanations of the meaning and role of urban movements in the 1960s and 1970s.

URBAN MOVEMENTS AND THE POLITICS OF LEFT-WING PARTIES

I shall start from a first observation. European countries which experienced the highest political mobilization in the 1960s and 1970s are at present the ones where urban movements appear to be almost nonexistent, and where there are no visible chances for them to boldly reemerge in the near future. This is the case in Italy, Spain, and France. At the same time, these countries have witnessed a rapid growth of political power of left-wing parties in local governments. In Italy and Spain, communist and socialist coalitions run most urban and regional governments. In France, the strength of the left in localities has been confirmed by the returns in the elections for the presidency and the Assemblée Nationale.

Does this imply the existence of a direct relationship between the political mobilization activated by urban social movements and the voting behavior in subsequent local elections? Urban movements, in spite of their conflictive attitude toward well-established left-wing parties (namely the Communist and Socialist parties), may have made voters aware of crucial issues in local government — the delivery of public services, housing, a new way of governing, and so on. They might also have brought into being loose but effective networks of political organization, which might in turn have funneled consensus toward less militant but still left-oriented forms of government. This deeper political consciousness of large electoral strata and an increased ability in political mobilization may later have translated into electoral support to left-wing parties.[2]

The cases of Italy and Spain fit well into a model in which urban social movements play an anticipatory role which can then produce a new political climate at the local level. But this is not the case in other European countries where urban movements have also been important. While it is possible to identify the influence of urban movements that were active in the 1960s and 1970s on the courses of action of left-wing political parties in subsequent years, there are substantial discrepancies in different countries in the way these relationships have taken place and in their outcomes. In West Germany, the interrelationship between the Social Democratic Party (SPD) and urban movements was more significant and fruitful in the late 1960s than in more recent years (apart from the very recent pacifist positions of large segments of the SPD, which are very close to the ones of pro-peace and anti-nuclear movements). And as far as I know, the massive political mobilization in Portugal in the mid 1970s has left little mark.

URBAN MOVEMENTS AND ECONOMIC PROCESSES

My second observation pertains to the relationship between urban movements and economic conditions. Struggles for housing, transportation, community services, lower costs of utilities, lower prices in basic consumption goods, and more generally, struggles connected with the process of reproduction of labor are often explained as the consequence of worsening structural conditions at the end of the growth cycle in the 1960s, and of the increasing demand for economic equity by newly emerging social groups (DaOlio, 1974). Yet the economic conditions of large social groups in several European countries are in real terms worse now than they were a decade ago. Protracted double-digit inflation has offset most of the gains lower-income groups among blue- and white-collar workers made in the 1960s and the early 1970s. Mechanisms of indexation for large segments of the core labor force do not compensate for the rising prices that confront the increasing numbers of unemployed, underemployed, and temporarily employed workers. It is particularly difficult for women and young people to find permanent and decently paid jobs. In addition, retired people and people on welfare are badly hurt by cuts in public expenditure.

Increased living costs coupled with a substantial reduction in public expenditure were expected to produce a kind of explosive mixture which would fuel social protest. This has not happened. Why? Once again, situations vary among European countries, and these differences have to be taken into account when an explanation is provided. In countries where urban social movements were very intense, a complex dual economy developed in the 1970s. While the formal sector has experienced a severe crisis, the informal economy has provided jobs, produced income, and at least partially balanced the decline in the formal economy. People have been progressively forced to adjust to this dual system — to look for temporary and semilegal jobs instead of permanent employment, or to be flexible enough to have several part-time jobs. Combining welfare revenue with an illegal job has become a current practice in large urban areas. Unions have reduced their control over these sectors of the labor force, and as a consequence, it has become more difficult to organize mass protest and to maintain an extended and permanent network of political activists.

The experience of Italy and Spain does not apply to West Germany, where social expenditure has continued to be very high in recent years, and where the economic policies of both the federal and regional governments have helped to contain the social protest. It does not apply to France either (Poulantzas, 1978). In the United Kingdom, a unique blend of welfare-state policies and individual devices to "muddle through" the economic crisis can probably account for the relatively circumscribed urban conflicts which have occurred so far (Gershuny and Pahl, 1980). But the Brixton riots are an indication that current solutions are not viable for the future (TUC, 1981). To complete the analysis of the relationship between urban movements, the party system, and economic processes, I want to introduce a further observation.

Changes experienced by urban movements in the past decade cannot be interpreted in terms of the political system only — i.e., the evolution of the political system and the different relationships which have developed between urban social movements and the political elites. It would also be misleading to assume that economic motivations play a too dominant role. The reductionist bias of theories which analyze urban social movements mostly in the light of the "cost of reproduction of the labor force" has proven to be a major weakness in their ability to explain real processes.[3] Political and economic components are strongly intertwined. Many of the aspects of the process of change are to be explained on the basis of their interaction.

INTERNAL SOURCES OF CHANGE IN SOCIAL MOVEMENTS

A third component must be taken into account to understand why urban social movements have modified in time and progressively faded out.

Participants in urban social movements in the 1960s and 1970s were fighting for lower fares, more participation in decision-making and better housing, but also for a society based on a different system of values, or at least different priorities among commonly shared values. Their different demands and interests combined together in a kind of "package" of goals (values) to be pursued and objectives to be reached. In time, both the ingredients in these packages and the relative weight of each ingredient changed. Objectives that were dominant at the end of the 1960s, for instance self-government, became less important in subsequent years. Issues such as transportation and housing, around which mass mobilization had taken place in the early 1970s, increasingly lost their relevance.

This process of change can be easily accounted for by several factors. First, actors changed (students and young blue-collar workers at the turn of the 1960s, women in the mid-1970s), and each leading group brought into the movements different goals and priorities. Second, as demands by urban movements were at least partly met and the political system adjusted in order to anticipate and solve problems, the focus of political mobilization shifted to new issues. Third, social groups progressively adjusted to the slowdown of economic growth, the fiscal crisis, the worsening of services, and so forth. Changes in expectations can also be found with reference to the issue of political representation. A new polarization seemed to take place. In several cases, a number of problems were assumed to be the responsibility of parties and other institutions, while other issues were considered to be crucial in terms of identity and personal life (Ergas: 1981a, 1981b).

THE CASE OF ITALY

A more detailed examination of some aspects of the political process which has developed in Italy helps to reveal the relationship between national politics, political parties, and urban movements. Italy seems to be the country where the most interesting political changes produced by the mass mobilization in the 1960s and 1970s can be traced.

During the past decade, this country experienced a very high level of political organization, and a variety of struggles which have no equal in other European countries. Most Italian cities witnessed extended social conflicts. Conflictive actions covered a wide range of issues: the occupation of vacant houses and rent strikes, struggles for better mass transit, struggles for direct control and management of community services such as health centers, schools, recreational facilities, self-reduction of telephone, electricity and other utility bills, and so on. These struggles involved almost every social group: young people, women, senior citizens, whole families of southern immigrants who did not have a decent place to live, middle-class commuters, blue-collar workers in the large industrial plants of Northern Italy (Marcelloni, 1979; Della Seta, 1978).

The political system, both locally and nationally, was forced to face the variety, size, and implications of such struggles. Following the mass political mobilization of those years, new left-wing groups were formed, while the traditional parties of the left (the PCI and the PSI) were forced to some degree to modify their strategies and tactics.[4] Even Catholic organizations were affected by these changes: New populist organizations emerged which were inspired by the ideals of equity, self-government, and participation which characterized that period. The unions were deeply involved in this process of change and often directly supported the struggles and the grassroots organizations that were behind them.

Two political developments have arisen from this original situation: absorption of popular demands into institutional politics, and radicalization of fringe elements.

ABSORPTION OF POPULAR DEMANDS

The two major parties of the left, the Communist and the Socialist parties, as well as the unions, have incorporated a large share of the pressure exerted from below. This has happened in two ways: (1) They have reorganized and adjusted their politics, in order to take at least partially into account the new issues expressed by grassroots movements — more democracy in decision-making, a deeper attention to the new problems of Italian society, and a lesser ideological dogmatism in the definition of politics (Lange, 1979; Tarrow, 1979). (2) At the same time, thanks to successful local and regional elections, they have been able to answer, at least at the local level, some of the demands that urban movements had made in the most conflict-laden years. This has partly delegitimated the grassroots movements. It has also fragmented their demands, and has forced such movements to become institutionalized in order to survive (Ceccarelli: 1981).

RADICALIZATION

The counterpart of this effort to redirect urban movements into the channels of traditional politics has been a radicalization of the extreme fringes of social movements. And this has in turn produced new and more violent conflicts, terrorism, and an increasing repression.

The main reason for the present state of conflict and repression can be easily pointed out. Italy is once again at a critical stage (probably the most critical so far) of the struggle for the political control of the state which began with the collapse of fascism in 1945. Since then, the Italian working class has increasingly won more power and rights, *but*, and this is crucial, it never has gained full access to the institutional system. Italy is still a country governed by parties representing middle-class interests and values. The judiciary is almost totally in the hands of the bourgeoisie. The army has been only marginally permeated by democratic values. Top state and parastate bureaucracies are still a self-contained elite. Parties which represent working- and lower-class interests (or are assumed to do this), besides having a strong grip on city and regional governments, have strong leverage on central government institutions, and strong influence on policy-making at the national level. But since they were expelled from the national government coalition in 1947, they have never formally participated in the governance of the country. This is not a minor handicap in the power game in a democratic state.

The attempt to prevent the working class from becoming a legitimized partner in the control of the basic institutions of the Italian Republic has been an antagonizing process. Italy is not a social-democratic society in which the working class has eventually been integrated and shares some measure of power with the dominant classes. Whenever it was possible for the bourgeoisie either to reestablish the relative hierarchy between social groups or to take back some of the power it had been forced to share with other social groups, it never hesitated to do so, be it by means of its power in the Parliament or by force.

It is worthwhile to examine how this conflict developed over the years. Three distinct stages can be identified:

(1) Post-fascist democracy was grounded on a basic assumption: Italy needed a modern capitalist economy, and henceforth a larger and more modern working class. This was necessary to enable the country to play the role the new international order had established at Yalta and Bretton Woods. A modern capitalist economy and a larger and more modern working class also required that a wider social and political "space" in the state be allowed for the workers. In the Italian situation, this new role was not just given by the bourgeoisie in power to the blue-collar workers and peasants; they had won it, given the crucial role they had played in the struggle against fascism and in the

establishment of the Republic. The emergence of a strong and well-organized Communist Party (PCI) as one of the major political parties in the postwar years had sanctioned this right to more political power to the working class. The PCI's strategy was essentially concerned with the creation and the strengthening of a modern working class.

The new democratic state was founded on this assumption, but the political space given to (seized by) the working class was not clearly defined. At the end of the 1940s, a political juncture favorable to the bourgeoisie occurred — the cold war and the rapid growth of a new middle class. The first major class confrontation took place — the Communists were thrown out of the government and, after the electoral debacle of the left in 1948, strongly antilabor policies were implemented. During the late 1940s and the 1950s the bourgeoisie pursued a basic aim — to prevent the working class from converting its contribution to the performance of the Italian economy into more political power. Italian peasants had to become industrial workers, and a large pool of urban workers was needed (in those years millions of people moved from southern Italy to the industrial areas of northern Italy), but this new working class had to be kept out of the centers of political power, and possibly deprived of its basic political rights.

The best and fastest way to reach this goal was to use sheer repression. The police force was overexpanded. Antiunion practices became a standard policy of most of the companies; the Communist party became increasingly isolated in the political arena. Wherever forms of political representation and organization of the working class existed, control and repression were enacted — even in local governments.

(2) This repressive strategy was defeated by the changes which took place in the Italian economy and society in the 1950s and early 1960s. During those years, Italy grew at a very rapid pace, and became a fully industrialized country. Very large urban areas had developed; modern patterns of consumption had become widespread; the role of the working class both as a producer and a consumer had become very important. Given this social and economic situation, it was increasingly difficult to deny more rights and political power to a class which played a crucial role in the society. In spite of violent repression, this class had been able to strengthen its political organization. In addition, after the long winter of the 1950s, the unions had reorganized, and they were again very militant. Even the capitalist camp showed changes. Conflicts between the most advanced capitalist groups connected with modern industries, the service economy, the banks, and the more traditional capitalist groups tied

to the rural world and the urban landowning developed. Capitalist interests were fragmented, and often in conflict. This change was expressed politically by the decline of the Christian Democratic power and the rise of new center-left government coalitions, including the socialists.

Center-left governments emphasized the role of the state in the economy, pressed for social reforms, and asked for the democratization of state institutions. In the political climate of the 1960s, a new consciousness of the political and social rights that a modern and industrialized democracy must guarantee to its members took shape.

Mass conflicts developed starting in 1968 which involved tens of thousands of students, factory workers, and consumers. They challenged the very nature of a system which aimed at the "containment" of rights of a large portion of Italian society rather than the guarantee of equal rights to every member of it. Institutions like political parties, elected bodies, and representative democracy were also challenged. The drive against all forms of representative power and authority, while reintroducing values that were at the very root of a democratic system, also produced a dangerous crisis of legitimation of bourgeois institutions that still had the role of protecting basic citizens' rights.

(3) When the spontaneous social and political movements began to lose their momentum in the mid-1970s, the proposal of a new political paradigm for the next decade began to take shape. In a very schematic and simplified way, one can say that two opposing strategies emerged. One was worked out by the biggest parties — including the Communist party — the unions, and important capitalist groups. It postulated that the organized working class had full citizenship in the state — equal rights to those of the middle class (in the process the boundaries between these classes have become blurred), and legitimate access to political decision-making. In exchange, they were expected to soften their economic and social demands (Tarrow 1979; Ceccarelli, 1981).

A few aspects of this project must be pointed out. It took place at a juncture of economic crisis and social instability. Marginal groups in the society and a large part of the working class did not fit in within this agreement. The coopted segment of the working class was expected to take care of these contradictions and mitigate the frictions and conflicts they originated. Features of neocorporatism were anticipated, and the agreement was worked out by the leaderships of the major parties. It was essentially a strategy elaborated and agreed upon at the top; mass support was built up later.[5]

Some groups of the left and some segments of the working class did not accept this social and political pact. Their strategy was to develop social conflict in an extreme form, in order to prevent this partial integration of the working class into the state. They postulated that if irreconcilable differences between the PCI and the Christian Democrats were maintained, and the split between their functional interests was kept open, the strategy of political compromise and of articulated and limited agreement between classes would inevitably fail. It is on these grounds that terrorism and repression developed in the late 1970s.

When urban movements reemerged temporarily in 1977 and 1978, they took place in a context which had changed dramatically. The solidarity which existed at least in part among urban movements, left-wing parties, and unions was gone forever. Two opposite camps confronted each other. In one, left-wing parties and the unions; in the other, radical fringe groups (the so-called *autonomia* and terrorists). Symbolic subjects of confrontation were "red Bologna," the model city of the Communist party, and the CGIL, the union controlled by communists and socialists. At the end of the 1970s, urban movements changed in character: one segment radicalized and increasingly alienated; the other increasingly absorbed into existing institutions. This affected the extent and the degree of political mobilization. Most of the traditional actors of social movements were uneasy with the polarization which had taken place. It was very difficult for them to either fully delegate the PCI and the PSI to represent their interests, or to mobilize around issues which were often dominated by quasi-terrorist organizations. There was a sharp decline in political participation, and political parties gained a considerable measure of control of grassroots, community-centered organizations. Mass mobilization without direct involvement of the major left-wing parties still took place around issues such as abortion, nuclear power plants, and nuclear war. However, mass movements of the late 1970s and early 1980s have little to do with the urban movements of ten years earlier. In addition, these movements are in most cases eager to build up alliances with parties in order to have a stronger impact on central government policies and on lobbying efforts for better legislation.

NEW ECONOMIC CONTEXT

During the 1970s the economic context had also changed substantially. Reproduction costs (one of the factors which had caused the conflicts at the end of the 1960s) had increased, while opportunities to find jobs had been sharply reduced. In addition, two other changes occurred:

(1) The demand for jobs had become more flexible, and workers no longer considered big companies in the formal sector of the

economy or the government as the only attractive employers. In the 1970s, industries in cities had undergone a substantial process of reorganization. Large-scale obsolete and noncompetitive plants either shut down or reorganized, and permanent jobs in large- and medium-sized plants were reduced. In the meantime, small industrial businesses flourished, creating a substantial amount of new jobs. A variety of firms and types of productive organizations then developed, from high technology small-scale plants, to cottage industries producing either traditional or modern manufactured goods. Consequently, in several urban areas the decrease in the number of permanent jobs in large industrial plants did not result in massive unemployment. The proliferation of modern small-scale plants, and the decentralization of production in artisan workshops and cottage industries, coupled with the increase in the supply of services, absorbed both the surpluses in the industrial labor force, and workers entering the labor market for the first time.

(2) Attitudes and expectations toward work have changed. Flexibility in employment and a certain amount of independence in work activities are considered highly desirable. Young people become acquainted with a mix of temporary jobs plus welfare as an acceptable form of living. A number of social groups, which in the 1960s and the 1970s had found in poor living conditions and the exclusion from basic rights a cause of protest, live now in different conditions. They have rearranged their expenditure patterns and can often afford higher costs for some services, even though they have lower real incomes. They request different services, and use them differently than in the past. The family structure and the conditions of women have to some extent changed, and this affects patterns of family expenditure and the use of social services (Ergas, 1981a; Maffii, 1981, Malgieri, 1981).

This process of change and adjustment does not mean that the causes of conflict have disappeared. But the whole present situation is less rigid than a decade ago, and there is little chance that urban social conflicts may explode again in the near future with the support of large social groups.

CONCLUSION

Rather than attempting to summarize and draw a conclusion as a result of this overview of the relations between urban movements and the political and economic system in Europe, I want to suggest three problems which ought to be further explored in future research:

(1) Urban movements in European countries in the 1960s and 1970s do have some common features, but when examined in detail

they appear to differ substantially in content and in the role they perform. This is due to the fact that they have taken place in very different contexts, which have evolved according to different dynamics. Differences instead of similarities ought to be stressed and better analyzed.

(2) Components of urban movements have changed according to different patterns and have combined differently over the years. This is the case also for similar economic problems, which, having taken place in different political situations, consequently have different meanings. Increases in mass transit fares — to give an example — have a different impact in cities controlled by left-wing governments than ten years ago when either conservative or moderate coalitions were in power.

(3) Goals and objectives of social movements vary over the years. The same can be said for values which are pursued, and for functional interests of individuals as well as of social groups. I think that all these variables must be taken into consideration in order to understand fully the nature and character of social movements.

NOTES

1. See on this the important collection of essays by Blackmer and Tarrow (1975).

2. Sivia Maffii and Patrizia Malgieri of the Politecnico di Milano have thoroughly analyzed how the conflict between urban movements and the city government in Milan has progressively turned into support for left-wing party coalitions in the course of the seventies (Maffii, 1981; Malgieri, 1981).

3. Economic reductionism characterizes most analyses by French students of social movements in the mid-1970s — Castells, Cherki, Godard et al. (Castells, 1978; and Castells, Cherki et al., 1974). This approach has been later submitted to self-criticism by the same students.

4. Peter Lange has done an excellent analysis of this process in his essay "The PCI and Possible Outcomes of Italy's Crisis" in Graziano and Tarrow (1979).

5. I have analyzed how this process has taken place in urban governments in my study of the political strategy of the PCI at the local level (Ceccarelli, 1981).

REFERENCES

BLACKMER, D.L.M. and S. TARROW [eds.] (1975) Communism in Italy and France. Princeton, NJ: Princeton University Press.

BORJA, J. (1975) Movimientos Sociale Urbanos. Buenos Aires: Ediciones SIPA-Planteos.

BOYTE, H. (1980) The Backyard Revolution. Philadelphia: Temple University Press.

CASTELLS, M. (1980) "Local governments, social change and political conflict." Political Power and Social Theory 2.

——— (1978a) "The citizens' movement in Madrid." International Journal of Urban and Regional Research 1.

——— (1978b) City, Class and Power. London: Macmillan.

——— (1977) Ciudad, democracia y socialism. Madrid: Siglo XXI.

——— E. CHERIE, F. GODARD, and D. MEHL (1974) Sociologie des Mouvements Sociaux Urbains. Enquete sur la Region Parisianne. Paris: Ecole de Hautes Etudes en Sciences Sociales.

CHERKI, E. and D. MEHL (1979) Le noveau embarras des Paris. Paris: Maspero.

CECCARELLI, P. (1981) "Local government control and the political strategy of European communist parties." Political Power and Social Theory 3.

——— (1979) Venice: urban renewal, community power structure and social conflict," in D. Appleyard (ed.) The Conservation of European Cities. Cambridge, MA: MIT Press.

DA OLIO, A. (1974) Le lotte per la casa in Italia. Milano: Feltrinelli.

DELLA PERGOLA, G. (1975) Citta e conflitto sociale. Milano: Feltrinelli.

——— (1974) Diritto alla citta e lotte urbane. Milano: Feltrinell.

DELLA STEA, P. (1978) "Notes on urban struggles in Italy." International Journal of Urban and Regional Research 2: 303-329.

ERGAS, Y. (1981a) " 'A republic founded on labour . . .' notes on women's citizenship in contemporary Italy." Presented at the American Political Science Association Annual Convention, New York.

——— (1981b) Feminism and the Italian Party System. (unpublished)

GERSHUNY I. and R. E. PAHL (1980) "Britain in the decade of the three economies." New Society (September).

GOUGH, I. (1979) The Political Economy of the Welfare State. London: Macmillan.

GRAZIANO, L. and S. TARROW [eds.] (1979) La Crisi Italiana. Turin: Einaudi.

LANGE, P. (1979) "Crisis and consent, change and compromise: dilemmas of Italian communism in the seventies." West European Politics 2, 3: 110-132.

MAFFII, S. (1981) Politiche e lotte per i trasporti urbani a Milano, 1968-1980. Milano: Istituto di Urbanistica del Politecnico. (mimeo)

MALGIERI, P. (1981) Lotte per la casa e politica del Commune di Milano, 1968-1980. Milano: Istituto di Urbanistica del Politecnico. (mimeo)

MARCELLONI, M. (1979) "Urban movements and political struggles in Italy." International Journal of Urban and Regional Research 2: 251-267.

POULANTZAS, N. (1978) Le pouvoir, l'état et le socialisme. Paris: PUF.

——— (1975) Classes in Contemporary Capitalism. London: New Left Books.

——— (1973) Political Power and Social Class. London: New Left Books.

PRETECEILLE, E. (1980) "Left-wing local governments and services policy in France." Presented at the CES, Harvard University Seminar on Southern European Urban Problems.

TARROW, S. (1979) "Italy: crisis, crises or transition?" West European Politics 3: 166-186.

TOWNSEND, P. (1979) Poverty in the United Kingdom. Harmondsworth: Viking.

Trades Union Congress [TUC] (1981) Regenerating Our Inner Cities. London: TUC.

13

Governmental Responses to Popular Movements: France and the United States

SOPHIE N. GENDROT

> We have run as far as possible, to the desert, to the heart of the forest, to the furthest point from everything. There, we have found the state.
>
> *B. Hervieu and D. Leger*

□ THE TRADITIONAL IMAGE of France in urban politics is that of a class-based society controlled by a centralized state with little citizen participation at the local level. Yet striking parallels have emerged in the last thirteen years between France and the United States in terms of the nature of popular movements and of governmental responses to them, notwithstanding sociohistorical differences between the two countries.

The aims of this chapter are:

(1) to describe the diversity of urban political movements in France since 1968 and to show the differences from and similarities to recent urban political movements in the United States; and

(2) to describe and explain the character of governmental reactions in France to these movements, again making comparisons with the American case.

AUTHOR'S NOTE: The author wishes to thank Steve Erie, Norman Fainstein and Joan Turner for their comments on earlier drafts of this chapter.

FRANCE AND THE UNITED STATES: TWO DIFFERENT NATIONAL SETTINGS

At first, it seems a challenge to compare countries whose institutional, social and political patterns are so different. In France, state power[1] is not fragmented by other governmental structures such as states, counties, cities, or neighborhoods. The weak parliament, the judicial system, and the media hardly check the French state, which holds the most extensive powers among Western democracies. In contrast, federalism, a system of checks and balances among powers, and jurisdictional fragmentation diminish the capacity of American public authorities to make decisions for which they are accountable.[2]

Economically, the French state is interventionist. To prevent businesses from resisting its decisions, it is willing to absorb the investment risks which would be taken by corporations in the United States. The relationship between the state and French businesses is cooperational but, as a counterpart, the economic market is subordinated to the political system and sometimes to arbitrary decisions *(le fait du Prince)*.[3]

While the local political system in the United States is the locus of contradictions between demands for capital accumulation and demands for social expenditures, the fiscal crises experienced by American cities are unknown in France. Until now, French cities could not collapse fiscally, since their budgetary balances are controlled by the Prefects (the local representatives of the central state) and ultimately approved by the Ministry of Finance.[4] Responsibilities for social amenities, i.e., schools, day-care centers, sports centers, are left to the initiatives of the cities. Yet, here again, cities follow budgetary directions established at the national level. Moreover, if deficits occur, the practice of *"péréquation"* comes into action — that is, shifting resources of economically active cities to those lacking resources. In contrast, extensive transfers aiming at redressing social injustices or uneven development would be interpreted in the United States as governmental intrusion in local and private affairs.

Differences in state structures are accompanied by contrasting traditions of citizen participation. It has been estimated, for instance, that in 1970 there existed 6 million voluntary organizations in the United States which involved about 70 million active volunteers (Smith et al., 1972). American voluntarism has given birth to organizational structures with access to the legislative and administrative bodies of government. The professionalization of volunteerism and

SOPHIE N. GENDROT   279

the ease with which Americans "get organized" to protect their interests have contributed to the institutionalization of popular participation and to its legitimacy. This phenomenon is increased by the relative weakness of fragmented local institutions and of political parties.

In contrast, French government and administration have been notably insensitive to outside citizen input. As a consequence, the 300,000 associations which flourish in France, until recently, rarely stepped into the domains reserved to the state. The notions of "turf" or "neighborhood" or "self-help," which have a powerful meaning in the United States, are weak in France. For a long time the French have been accustomed to seeing their affairs settled by distant decision-makers. As Tocqueville says, *"Pour eux, le gouvernement a trop pris la place de la Providence."* Yet it is important, here, to emphasize a factor which has been frequently overlooked in studies on French centralization, that is, the national consensus on representative democracy. Why is the level of participation in local elections so high in France? Besides civic culture, it could be that a structural element called *"le cumul des mandats"* contributes to explain such behavior. As the French system allows cumulative elected positions (i.e., a mayor can also be a member of the Parliament), local officials are given the power to negotiate the interests of their constituents at various levels and from various perspectives. A French sociologist accurately noted that French local officials perform both a role of representation and the function of loosening the rigidity of centralization. Local power is held by elected officials and street-level bureaucrats and not by local organizations, which are still held in suspicion by the local "establishment" (Grémion, 1978).

The real contest for the French state and for the classes it represents has come from the political parties of the left and from the powerful trade union confederations. This situation has no real equivalent in the United States. The left institutional forces embody the working class and part of the middle classes. They hold power at the local level in many regions of France and their representatives at the national level act as "watchdogs" on government policies. The very small margin separating the left from the right electorate forces the state to provide numerous social transfers to the benefit of the population. As a consequence of capitalist restructuring — what Joseph Schumpeter would call creative destruction — welfare problems, which in the United States are left to the care of deficit-ridden local governments, are treated nationally in France. It is understood that "if the twisted logic of capitalist development creates masses of human debris, . . . the government has to find a way to take care of these victims of creative destruction." (Cohen and Goldfinger, 1975: 77). In contrast, both social and geographical mobility in the United States have enabled the more fortunate to migrate from social problems. They have also prevented, among other things, the development of strong leftist institutions. In the multiethnic, multiracial American society (as opposed to the more homogeneous and historically rooted French society) the reality of class has not been openly acknowledged or dealt with.

RECENT CHANGES IN FRENCH AND AMERICAN SOCIAL LANDSCAPES

The May 1968 movement in France, stemming from a students' revolt, triggered a general social and political crisis. After 1968, cracks appeared in France's powerful centralized model. As in the United States, a crisis developed in the decision-making system, in social relations, in values and culture. Complex situations, such as the restructuring of the steel industry (with massive layoffs) or the development of public works (with mass displacement) or the control of public transportation costs (through fare increases) were now resolved with increasing difficulties. France's economic modernization was paralleled with a weakening of authoritative structures — of political institutions as well as of integrative structures like school and family. People still expected the state to provide for social assistance, security, justice, better living, and stability. They were proud France kept up with the advanced industrial societies. But on the other hand, as domestic détente went on, there was a widespread view that the administration had become too bureaucratic and ineffective to deal with the special problems of a complex society. After 1968, as in the United States, a general questioning of modes of social behavior (practices) took place in France. It touched a great number of groups and sectors. For example, a women's liberation movement, and regional minorities movements developed, especially in Brittany and Corsica. Practices in the prisons and psychiatric institutions were challenged, as well as in the police and in the judicial system. Simultaneously, urban protest movements denounced the isolation of bedroom communities, profit-oriented urban renewal operations, discriminatory housing policies toward immigrant workers, and poor transportation. Such denunciations and resistances to patterns of capitalist accumulation and to state deficiencies in the field of collective consumption were new in France.

Simultaneously, there was a general disaffection toward political parties and representative democracy. May 1968 — direct expression and participation in the street — had particularly revealed the crisis of older party structures and their incapacity to meet new demands. It created a vacuum which grassroots organizations attempted to fill. Additionally, as economic growth proceeded unevenly and unemployment grew, the labor movement became stronger and reached new categories of workers among the young, the regional minorities, and the immigrant workers. New forms of struggle appeared such as worker-controlled Lip in the clock industry or Joint Français in Brittany.[5] As state intervention increased in the production, distribution, and management of such collective goods and services as housing, education, health, transportation, and mass media, frustrations grew. Aspirations for communal life, for local autonomy and expression, for leisure, challenged the bureaucratized patterns dealing with *métro-boulot-dodo* (subway-work-sleep). Along with social unrest and unemployment, a concomitant crisis in values and expectations undermined traditional capitalistic values.

During the same decade, in the United States, a political, social, and ideological crisis threatened the stability of advanced capitalistic governmental structures. After the civil rights movement, the ghetto uprisings and the student riots, social unrest did not disappear: Municipal workers' strikes, neighborhood movements, tenants' strikes, resistances to evictions and displacement, and consumers' struggles disrupted the prevailing social order. Mobilized groups attempted to resist bureaucratic intervention, planned shrinkage of services, and the all too visible hand of the market.

These similarities allow, therefore, a comparison of such unrest in both France and in the United States. For convenience, we label these collective forms of protest and mobilization arising from urban contradictions "movements." In fact, except in rare cases, these stirrings of discontented groups do not produce social change. They are rather learning processes to politicize and mobilize the victims of economic crises and state power.

POPULAR MOVEMENTS IN FRANCE AND THE UNITED STATES: THE COMMON CHARACTERISTICS

Traditional class struggles cannot help us understand the new social struggles. Their organizations and processes differ — from confrontation to cooperation or a blend of both — and so do their outcomes. A common denominator between these movements is elusive. In both countries — and these are new features for France — the struggles are fragmented, locally based, pluriclassist, and weakly structured. Some of them are defensive (mobilization of an area against an external threat) or active (local mobilizations around a free, nongovernmental local radio, in France). Some deal with a specific issue (housing); others fight on several fronts (Association of Community Organizations for Reform Now, ACORN). Some will last a few months (transportation issues), others for years (ecological issues). Some expect major social changes, others a conservative status quo.

Through a few examples of struggles in France and, whenever possible, of their counterparts in the United States, some situational factors explanatory of such struggles will be offered, and some national differences emphasized. Cases will draw more heavily on France, which is less familiar to English-speaking readers.

TRANSPORTATION ISSUES

The struggle around the transportation issues represents the first and the most spectacular form of a global and united opposition on an urban theme from Parisian users (Cherki and Mehl, 1979: 17). In Paris, 80 percent of all commuting trips and 70 percent of all trips are made on public transportation[7] (Hartman, 1980: 2). The system is controlled by the French central state through *Régie autonomes des transports parisiens* (RATP). Parisian mass transit fills such a crucial function in regulating regional development and complex labor and consumer markets that it has always been taken for granted that its deficit ought to be borne by government. By 1970, however, passenger fares covered only half the cost of maintaining public transportation, and the French national government supported 70 percent of the operating deficit (Heidenheimer et al, 1975: 170).[7] Following a large raise in 1967, a 16 percent boost was announced in the fare of Paris subways and buses in 1970.

Direct action was taken immediately by two small extremist parties (Parti Socialiste Unifié — PSU — and Lutte Ouvrière) to organize a *fédération* of local users' committees. The 120 local committees' actions denounced the deficient organization of the transportation network (lack of comfort, frequency, and coverage) and particularly the lack of linkages among suburbs detrimental to the working-class population.

Simultaneously, a vast campaign was launched by all the leftist political parties, and trade unions mobilized for the first time on an urban issue in France. Three massive demonstrations (20,000-50,000 persons) opposed the fare rise and demanded free transportation for workers regardless of distance. Such actions[8] brought victories to the two parallel movements. The government was impressed with the opposition and strength of the working-class transit users. Beyond the apparent object of struggle, a contest for power took place which forced the government to back down.

Such a phenomenon has no real parallel in the United States. Car owners there constitute a more powerful and numerous constituency than public transportation users.[10] Structural factors such as the institutionalized fragmentation of decision-making in the mass transit system, the segmentation of interest groups defending the users, and their reticence towards trade unions[11] may explain why a centralized protest movement has not yet taken place on transportation issues in the United States.

ENVIRONMENTAL ISSUES

After the ecology movement broke out in France in 1973-1974, specific transportation demands were diluted into other environmental concerns. Public resistance to highway constructions in suburban Paris or within Paris neighborhoods resulted in spectacular struggles and victories. Although most associations claimed a strict nonalignment with existing political parties and trade unions, they were at the core of the local political debate. During the 1977 municipal elections, all parties took local positions on the controversial highway constructions. Individually, members of these associations were candidates either from the Green party or the left or the majority parties. To preserve their unity, associations maintained an attitude of neutrality toward parties; at the same time, relations existed between them, and in some cases parties at the local level extended the movement's action (Cherki and Mehl, 1979: 145). The outcomes of the municipal elections in all the communities mobilized over the highway issue were directly influenced by the movements' actions.

Similar movements occurred in large American cities to oppose the bulldozing and dismemberment of communities. Pressures were put on mayors, who, facing reelection, would abandon too-controversial proposals. In 1967, neighborhood activists were able to press Mayor Lindsay to kill the Lower Manhattan Expressway. But in the 1980s there are new circumstances. The need for construction jobs and for federal dollars to reduce the city's budgetary problems, a new profit-oriented urban philosophy, and the personality of Mayor Koch explain why the West Side militants have been unable, so far, to trade

Westway for improvements in mass transit. As Yago (1981: 22) points out: "The United States remains the only western country which cuts its transit spending while increasing highway building. Every attempt at revitalizing mass transportation in the U.S. has been sabotaged by the industrial interests tied to highway transportation."

HOUSING ISSUES

Dramatic upsurges in the housing movements in the 1970s were witnessed throughout the advanced capitalist countries. Refusal to pay full rents or maintenance charges, rent strikes, squatters' movements, and collective resistances to evictions have confronted local or national governments, landlords and the real-estate industry.

Historically, as they were the least able to provide a stable and adequate living environment for themselves, tenants have been most active and militant in France as well as in the United States. Struggles on housing conditions were historically secondary fields of opposition linked to broader working-class struggles. What emerges now are new movements in the reproduction field aiming at building strong coalitions of opposition to capitalistic practices; in the same manner, trade unions fight for improved working conditions in the work place. Struggles are locally based and fragmented. In both France and the United States patterns of segregated housing and diversified, class-based housing markets split tenants' coalitions. Racism defeats multiracial, working-class solidarity. Ethnicity and race, which cement alliances, are established on class interests. Racism and ethnic antagonisms activated by diverging economic stakes splinter the impact of community coalitions (Gendrot and Turner, 1982).

In France, rent strikes in public housing to protest rent increases and maintenance charges have largely been middle-class movements, while immigrant workers protesting their living conditions led their own strikes in the segregated, semipublic hostels to which they are confined. Efforts by middle-class organizations to reach working-class tenants have often met with failure in French public housing. (Working-class families often consider their settling into bedroom communities less as an expulsion from the inner-city than as a concrete gain of improved housing). In the United States, many movements dealt more with the particular interests of homeowners than with problems common to all white or minority tenants, such as rent control or neighborhood stabilization. Yet, only when masses of people, like the 50,000 Co-op City rent strikers, came out on housing issues could they reach victory.

Tenants' movements in the Parisian region — e.g., Sarcelles and Val d'Yerres — which fought against rent increases in the 1960s were not as well organized. Residents — blue- and white-collar workers — had settled in the projects at different periods of time. Their networks hardly overlapped. As in Co-op City, tenants opposed a public developer, the Société centrale immobiliére de la Caisse des dépôts et consignations (SCIC). But the mobilization and solidarity of the rent strikers was in no way comparable with the Co-op City organization. Fractionalism among associations (some of them close to left political parties) of tenants and homeowners, and the lack of collective structures to control the strikes, were negative elements which would not break the dominant forces' determination.[12]

Moreover, the stakes, besides profitability of the program, were crucial for SCIC. Because of centralized structures in France, if SCIC, the most important developer of public housing in France, agreed to modify a local regulation relative to rents, the change would automatically affect every project currently managed (or to be managed in the future) by SCIC all over France. These elements explain why three residents' rent strikes, which took place in 1965 and 1971 in Sarcelles and in 1972-1973 in Val d'Yerres, were strongly resisted and had relatively few positive outcomes.

Squatters' movements. Less spectacular and permanent than squatters' movements in Berlin, Amsterdam, London, Copenhagen, or Milan, the French squatters' movement which spread in 1972-1973 in Paris and the suburbs presents its own originality, though the reactions it provoked can be compared with those of a similar movement in Philadelphia in 1977. The French squatters were not only young people as in Germany or the Netherlands. They were also immigrants and unemployed workers, craftsmen, and "marginal" elements. The North Philadelphia squatters were poor urban blacks. In both cases, squatting came from a lack of viable options: a lack of resources, a shrinking rental market, and bureaucratic constraints, leading the squatters to denounce what M. Foucault calls *"l'enfermement"* (the enclosure) to which they were structurally condemned.

An extreme-left grouping, Secours Rouge, provided the "enduring organization" needed to coordinate the various groups of resistance and give their action visibility. The intellectuals, mostly students from Secours Rouge, helped the squatters when they were summoned to court or when evicted. Petitions were written and street demonstrations organized. A central demonstration gathered several thousand persons. For Secours Rouge, the movement was a means of establishing a political base in working-class neighborhoods, and of developing a radical political offensive against the central state apparatus. They led a political struggle grounded on a service issue and not an issue-specific struggle with a political impact (Castells et al., 1978: 536).

In North Philadelphia, in contrast, Milton Street orchestrated what he called the "walk-in urban homesteaders" actions. Although Street was only one man, the press was his ally and his movement was given a large coverage. As E. E. Schattschneider once pointed out: "The outcome of all conflict is determined by the *extent* to which the audience becomes involved in it. That is, the outcome of all conflict is determined by the *scope* of its contagion" (1960: 2). Consequently, for the first months the federal authorities were sympathetic, even supportive of Street's program.

In fact, there was legitimacy in the squatters demands, despite the illegal character of their actions. The North Philadelphia program was a desperate solution to a drastic housing crisis. Heavy vandalism threatened the 10,000 vacant homes of the area (Philadelphia Tribune, November 22, 1977). Most of the occupied houses were the property of the U.S. Department of Housing and Urban Development (HUD). No rehabilitation program had been provided for the city's biggest slum by Mayor Rizzo's administration. In actuality, it would seem that the administration had long-range plans for that neighborhood and that the displacement of the residents would occur as a consequence of city-aided "gentrification" — the encouragement of the middle class to return to the inner city — land redevelopment.

In the French case, sparse coverage of Secours Rouge and the squatters' actions was given by the media, which were often self-censored. Little was said about the 250,000 vacant dwelling units in the Paris region — 8 percent of the housing stock (Castells et al, 1978: 415) — and about empty public housing (HLM) Units.[13] The scandal of speculation, in which the state openly took part through urban renewal operations (Godard et al, 1973), was little denounced. Yet, it was linked to the squatters' occupations of newly constructed public properties (though not exclusively).

The critical role performed by the intellectuals of an extremist party in such a movement is essential: the militants are needed to build a social force. Yet, their presence overpoliticizes the movement, and the squatters' specific mobilization is "detoured." Due to political cleavages, the French working class and the leftist political parties would not support them, even though the housing crisis was a mobilizing theme. Besides, the illegal character of squatting discouraged working-class support in France. (In contrast, Italian trade unions were often involved in campaigns to occupy housing, stop paying telephone bills and fare increases in public transportation, and reduce rent increases). The lack of political and media support, the revolutionary idealism of the militants, and the size of the stakes contribute to explain why, in this first type of a relatively radical and massive urban struggle, harsh retaliation was exerted by the central state apparatus and no legitimacy given to grievances.

Urban renewal. Related to the housing issue but far beyond it, urban renewal provoked violent resistances in French and American cities. In the United States, low-income residents, most of whom were nonwhite, and small businesses were displaced and often relocated in overcrowded and already deteriorating areas. To prevent the dislocation and the rebuilding of their neighborhood into an economically profitable and white area, defense organizations attempted to block the bulldozers. Such struggles have been difficult. Racism and fragmentation of interests divide the working class. Bias in the system, that is "the absence of neutrality on the part of official decision makers *and* the absence of neutrality in the institutions, norms, practices that encase them" (Stone, 1980: 6) put the lower strata in a systemically disadvantaged position.

Similarly, economic factors and public authority produced the same results in France. The large deindustrialization of Paris was followed by capital restructuring and the conquest of working-class neighborhoods (*Ve, XIIIe, XVe "arrondissements,"* Les Halles, le Marais, Belleville) by the state and by private developers. Deportation of working-class populations to distant suburbs provoked opposition. These struggles revealed wide urban contradictions and questioned the planning of urban growth. They indicated a primary importance to the reproductive sector as well as to the role performed by the state in the consumption field and in the regulation of social relations (Cherki and Mehl, 1979: 194). Yet, due to the fragmentation of the struggles, to their localism and isolation (especially when supported by extreme-left militants), opposition did not trigger an urban movement. Left political forces and trade unions did not mobilize on a stake which was decisive for dominant classes; and little pressure could be exercised at the local level since operations had been conceived and decided at a higher state level (Castells et al., 1978).

THE PROTAGONISTS

Middle-class reformers[14] have been heavily involved in urban struggles in both countries. They benefited for a long time from the consumption model, then they became the victims of inflation, of economic austerity, and of processes they could not control at the working place (subtle work divisions, task division, labor polarizations). In reaction, they try to regain power in the urban field. In leading the resistance to state policies or capitalistic interests, they aim at becoming the privileged agents of negotiation with the authorities, *within* the existing political framework. The most militant of them often have a political past: They fought for civil rights or against the Vietnam war in the United States; they were involved in the left or extreme-left parties in France and were disappointed by the whole process. They now defend their environment, their neighborhood, and the quality of their lives in opposition to the production-consumption approach which is supported by the traditional political parties.

Groups defending their cultural and ethnic identity and organized on a territorial basis are also participants in struggles confronting state policies. In the United States, but also in France, certain groups share a sense of cultural uniqueness. "While evidence is difficult to provide," Suzanne Berger (1977: 165) notes, "There seems to be, in France at least, a symbiotic relationship between regionalism and ethnicity." Such ethnoregionalist movements are linked to wide regional disparities in benefits, in the provision of public services, and in capital investment. Poorer regions with new weak industries experience more directly the impact of economic crisis, budgetary cutbacks, and declining investment. The new phenomenon in France, but also in neighboring countries, is that groups choose to contest the capitalist expansion from which they do not benefit, not as workers or peasants or government employees but as Britons or Corsicans.

Compared with ethnic mobilizations and conflicts in the United States, the French cultural demonstrations — highly visible when they resort to terrorism to protest against state policies — remain limited and they have not yet mobilized public opinion.[15] But connections with other groups are apparent. For instance, "The role of environmentalists in the evolution of the Occitanian movement over the Larzac case, involving the expropriation of land by the army for a military camp, has been important" (Berger: 174).

Working-class groups, including third world people, intervene to defend their material conditions of existence and in some cases, their mere survival. Their demands are mainly economic, as witness the struggles against tenants' evictions or the planned shrinkage of services. When we turn to the American mode of organization, a distinction must be made between community organizations fulfilling

service-delivery and populist, politicized, local, or coalitional movements. In our view, service-delivery organizations, publicly funded,
constitute less a movement than an imposed pattern representative of
social organization in the United States, to compensate for institutional deficiencies in providing social protection.

In France, the working class has not been involved as such into
urban struggles, but alliances have been formed with groups sharing
consumption problems at the neighborhood level. Immigrant workers
have been particularly involved in resistances to neighborhood displacement and in the squatters' movement in Paris. This pluri-
classism, however, is deceptive. The working class withdrawal from
urban struggles, Mehl perceptively notes, has certainly a social and
cultural background. But it also results from a suspicion toward urban
struggles, which leftist parties maintained for a long time as a consequence of a long tradition of distrust toward the middle classes
(1980: 43).

The opposition to governmental plans, especially in the services
and housing arenas, sometimes carry racist or class-interest
overtones. In the U.S., the antibusing movement, the opposition to
minority neighborhoods' control of local schools, to public housing
constructions by the white working- or middle-classes are examples
of this form of reaction. Racist attitudes of this kind were recently
exemplified in France, as, in the Presidential campaign, the electoral
strategy of the Communist Party was to activate the French
working-class in communist suburbs against immigrant Maghrebin
families over the local allocation of public funds to the latter's specific
needs.[16]

Citizen action groups. In France as in the United States, local
associations, tenants' associations, transportation users' committees, ad hoc committees are frequently linked with broader organizations which attempt to coordinate these manifestations of discontent.
Networks of citizen action groups like Fair Share, the Citizens'
Action League, and ACORN exist throughout the U.S. and have
built stable, large-scale organizations of protest to help the public
resist governmental and corporate policies. Efforts to overcome
fragmentation and localism (as expressed by the National Tenants
Union) prove to be the best strategy because adversary interests are
highly organized at the national level and can directly influence federal policies through lobbying and other means (Angotti, 1981: 32).

In France, the Confédération Syndicale du Cadre de Vie
(CSCV), Confédération Nationale du Logement (CNL), Fédération
des Usagers des Transports en Région Parisienne, though issue-
oriented, are organized at the national level. CSCV and CNL are

closely linked to left political parties. National and local organizations coexist and struggle at various levels and with different modes of intervention.

Such an outline, which only intends to point out that state deficiencies and capitalistic expansion in both countries mobilize fairly similar social segments, obviously has its limitations. Sociohistoric national factors, such as race and class, are important variables which differentiate the style of citizen movements in both countries and the governmental responses they receive.

GOVERNMENTAL RESPONSES TO POPULAR MOVEMENTS IN FRANCE AND THE UNITED STATES

Because there is no historical tradition to deal with movements which do not define themselves in terms of class and which are fragmented, governmental responses in France tend to look like the American patterns. A multitude of experimental measures are launched in order to adjust to specific cases. Although France is not as open to social innovation and experimentation as the United States, in the last ten years more flexibility and domestic détente have nevertheless emerged, and this trend has been recently accelerated by the change of government. As society is less "stalled" and as authoritative structures have weakened, special attention has been given to these new movements.

Governmental responses vary from integration and accommodation to displacement and neglect, as well, of course, as repression. Because repression is costly and may shake the elites' legitimacy (when it is a long-range method applied to large sections of the ruled), in terms of rapid social and economic changes democratic governments have resorted to accommodations, whenever possible.

INTEGRATION

France has not yet attained the flexible capacity of U.S. institutions to channel and orient participation. In the United States, during the 60s, federally funded programs (antipoverty agencies, Model Cities) were aimed at dampening social unrest. These "new machines," according to Katznelson, represented "parallel political institutions for the discontented." Minority leadership to a large extent was absorbed by such "machines," leaving increasing masses of disenfranchised people to themselves. In New York City, for example, attempts at institutional decentralization to respond to the lack of accountability of bureaucracies and to the demands of community control, were also made with the creation of little city halls, elected school boards, community planning boards, and area policy boards. Yet, participation in neighborhood institutions, far from remedying the situation of lack of representation, was a mechanism to mediate tensions and to divert lower-income groups from protest activities more useful to their cause. Once the goal of reducing unrest was reached, that is, as the mobilization effects decreased and as "law and order" prevailed, such programs were dismantled and their experimental policies terminated.

A new form of control has appeared in the 70s, diverting advocacy, confrontation-oriented organizations into neighborhood revitalization and economic concerns. Due to legislative requirements, vertical linkages between city or federal agencies and local organizations have developed and discouraged multidimensional horizontal networks. For lack of a wider political coalition to which they can relate, many local organizations have become overly technical and preoccupied with bureaucratic concerns. In the meanwhile, as the book says, "The rich become richer and the poor learn how to write grant proposals." As in the Community Development Block Grant example, contesting associations tend to become community development corporations (Fainstein and Fainstein, 1980) and bargaining becomes a substitute for confrontation.

The evolution from advocacy to mandated groups (Gittell: 1981), marking the limits of politicized participation, occurred also in France. As mobilized groups, especially middle class reformists struggling over environmental issues, have kept their distances from the traditional parties, collaboration with the state has been offered, emphasizing technical efficiency. Thus, the group's power to challenge the state has been neutralized. According to the predictions of a French sociologist, as the central state needs "new relays to stablize a social environment that traditional local elites no longer control," the mobilized middle class may become "the extension of the corporatist rationalizing State" (Grémion, 1978).

Another form of integration has been performed through the political institutions in France. Examples of local ex-activists running for elected positions are numerous in the United States. In 1977, in France, most parties, including the right, had incorporated ecologist, women's issues, and ethnoregional concerns in their platforms during the municipal campaigns. To preserve their "raison d'être" and get more leverage, some environmentalist leaders have been forced to run for elections as candidates independent from the traditional parties. Such a move, however, carries risks: (1) It divides the movement; and (2) when elected, these leaders lose their ability to mobilize a following and become detached from their constituencies. But if the organization tries to resist parliamentarism, it is also in danger. As institutionalized parties readjust their platforms to incorporate the new issues, they may attract the most militant members and provoke the disintegration of the organization. Such was the case of the Groupes d'action municipale (GAM) in 1974.[17]

How do governmental responses vary in relation to the different variables intervening in the movements, e.g., degree of mobilization, importance of the stake, public support given to the conflict, force of the organization(s), political context?

Concessions are more easily granted to large, organized, pluri-classist movements, such as the 50,000 Co-op City tenants in the Bronx or the 30,000 transportation demonstrators in Paris. Whether such concessions on large stakes are interpreted as tokenism or as evidence of political institutions' capacity to reform, they are not offered readily by governments. (Piven and Cloward, 1979: 29). In the case of Co-op City, confronted by the largest tenants' strike ever brought in this country, the state acted through its judicial power. State housing officials were determined to break the strike which was a threat to the profitability of the Mitchell-Lama programs. The strike steering committee was hauled into court, its leaders convicted. Bank accounts were seized, leaders' relatives and supporters harrassed and threatened with jail. Yet, the tenants' high mobilization and cohesion, their solidarity, were important assets, just as were the political connections middle-class leaders had with Democratic reformers and the voting strength the Co-op City tenants could exert. Democratic machine candidates, whose influence on the courts was determinant, were defeated in the primary elections of 1976, as a result of the alliance rent strikers made with Democratic reformers (Newfield and DuBrul, 1977: 310). The change of political context brought victory to the rent strikers. After the settlement with the state took place, Co-op City was directly managed by the tenants. This settlement, which may appear as a loss of control by the state and a victory of grassroots democracy, can also be interpreted as a simplification of bureaucratic procedures and the recognition that public management in Co-op City was a synonym for financial waste and corruption.

In the case of the Parisian public transportation users, their ambiguous victory took place at four levels: (1) The users did force the French government to back down three times from a fare increase. Though the fare was raised in the end, in August 1971 — a typical timing for political "nasties" — there was no more increase until 1975. The financial contribution of users to the costs decreased, whereas the government had to shoulder the additional deficit due to the absence of fare increases. (2) A payroll tax called *taxe d'équipement* was levied on Parisian businesses to finance mass transit.[18] This tax brought controversies: its passage demonstrated national political "muscle." In all likelihood, such a tax would have been opposed by corporations in the United States, and they would have threatened to leave the city. Central Paris remained too attractive to businesses for such threat to have any impact (Heidenheimer, 1975: 171). It is also easier for French elites, when decisions are taken at the national level, to have them implemented at the local level than it would be in the United States. (3) A monthly pass for unlimited rides on buses and subways[19] was introduced as a direct outcome of the 1970-1971 demonstrations, and the number of transit riders then markedly increased. (4) Another effect of the demonstrations and local committees' pressures led to the improvement of quality in the conditions of transportation.[20]

Thus, it appears that governmental responses to the transportation movements were positive because the balance of forces was favorable to the users. Exceptional circumstances — the aftermath of May 1968 and the union of the leftist forces — contribute to explain this victory but also its limitations. The grassroots movement gained strength, because at the central level the cartel of national parties and unions organized demonstrations receiving public support (Cherki and Mehl, 1979: 41). The Communist party and the Socialist party had an interest in picking the transportation issue. They demonstrated their capacity to organize massive demonstrations and to offer realistic and precise reforms. But such an alliance has its weaknesses: when the parties and unions in 1973 turned to other questions, the impetus of the movements soon faded away. Little had been accomplished to establish processing of demands by the local structures of the Fédération (FCU), and membership declined rapidly. Later on, action to protest the increases of the monthly pass price had little impact on governmental action.

One should not get the impression that the momentum of the movements fostered innovation. In fact, the movements sped up a restructuring which benefited productivity. The movements were unable to impose alternative policies, which would have dealt, for instance, with car policy in Paris.

Certain demands cannot be accommodated because they are too costly or revolutionary. On the walls of Paris in 1968 students had written: "Be realistic, ask for the impossible." When the squatters in Paris or Philadelphia occupied hundreds of apartments or houses, the stakes were too high and the demands could not be accommodated as such. As the disrupted mechanisms were central to the governments, efforts were made to delegitimize and *discredit* the activists. It was easier for the French government to present the squatters as deviants, anarchists, or revolutionaries because the "scope of contagion" was reduced, and extreme-left intellectuals supporting the squatters were politically isolated. Although the counter-propaganda in some cases reinforced the activism of those under attack, it usually delayed the mobilization of potential supporters. In North Philadelphia, the scope of the housing crisis and the support given by the public to Street's program induced HUD authorities to negotiate, in the beginning. But local pressures from the Rizzo administration and homeowners, implying that acceptance of the program could have consequences in other cities, led HUD officials to break the movement and repression followed. (As in a traditional American urban fairytale, a few years later, the squatters' leader, Milton Street, was coopted with a Congressional seat.)

Displacement and the segmentation of conflicts are methods used by governments to disrupt coalitions or to prevent movements from becoming centralized. The elites act on social cleavages and especially on racism. In the United States, the sense of ethnic consciousness is used — and sometimes created — to provide authorities with manageable, manipulable constituent links (Katznelson, 1973: 477). It could be argued that society is divided and that the state only reflects society; but to an important degree, the state has an autonomous and mobilizing effect of its own, reflecting its institutionalized values and its interest groups (Etzioni, 1968: 415). When unequal concessions are offered to residents evicted from their neighborhoods, whether they are minorities (or immigrant workers in France) or not, workers or middle-class, single or families, elderly or young, governments encourage what Yates calls "street-fighting pluralism" (Yates, 1977). In the most heated political battles, groups end up fighting among themselves rather than city hall (Jones, 1979). The manipulation of individual rewards and threats combined with the difficulty for heterogeneous working-class neighborhoods to maintain long-time lobbies impel changes in the attitudes and behavior of the militants.

If mobilization persists, however, and threatens capitalist development or affects a central institution, *repression* intervenes, because there is a social consensus to support repression. According to political and economic circumstances, the base of support is large or more restricted. As conservative societies, France and the United States have in common a long tradition of resorting to surveillance and repression. Until recently, France was labeled a "police state," especially by its second-class citizens. But there have not been, as in the United States, special target groups thrown to vindictive crowds and bearing the blame for domestic problems such as the striking workers in the 30s, the communists in the 50s, the black nationalists or radical students in the 60s, and now victims of economic oppression or state intrusion. We could hypothesize that repression, "the seamy side of democracy" (Wolfe: 1978), has marked the history of the United States and that mechanisms of integration developed in the 60s and 70s have been more of a parenthesis, an exceptional period when conservatism has yielded to popular support for change. Anti-left attitudes in the public at large encourage, in general, strong sentiments in favor of official violence (Gittell, 1981: 6). This support is increased in periods of economic instability. Harassment, arrests, evictions victimize minority groups, extremist organizations and associations whose dissenting ideas do not fit into the dominant ideology. As Wolfe points out "from the least violent form of repression to the most, the class bias of the state is the most important factor in determining the severity of the means" (1978: 98). Filipino tenants in San Francisco, neighborhood groups opposing the closing of Sydenham hospital in Harlem, older people in the single-room-occupancy hotels will be more harshly repressed than white, middle-class groups. Symptoms of apathy perceived in groups subject to rent or public transportation fare increases or victims of planned shrinkage of services or of racist policies should be interpreted against the background of police repression.

Besides a conservative ideology crisscrossing social classes, another important factor is offered by the political environment at the local, but most of all, at the federal level, which encourages or discourages official and unofficial repression. Contrasts prevail between the Reagan and the Mitterand administrations' attitudes toward demands for change. Revolts against budgetary cutbacks, no longer supported by a spirit of liberalism, are bound to trigger harsh retaliation in the 80s.

There is no intention to argue that the new Socialist government in France will not resort to the methods of repression used by the previous government against squatters, immigrant workers, or extreme-left militants formulating radical demands. But, as there is now an impetus *within* the system for social reforms, provided by militants from the left, and a wide popular support to give them legitimacy, it is likely that repression of demands will intervene only as the last resort. As an example, ethnoregionalist groups whose grievances were not channeled into the system were imprisoned and considered as outlaws by the previous government. In the same manner, antinuclear militants were punished, as they opposed state policies. The first steps taken by the new government have been to launch vast national debates on decentralization and the roles regions should perform, and on the nuclear question. The Parliament is also studying bills which would grant tenants their rights and associations their status. Whether these reforms are tokenism or not, whether the government will be able to meet all the expectations created by the politics of social change, it is too early to say. But the new political style seems to encourage the integration of demands rather than their repression.

CONCLUSION

The cross-national analysis has pointed out that similar structural contradictions in the development of advanced capitalist countries trigger similar social processes. Notwithstanding different political, social, and physical settings, large categories of social segments tend to mobilize on new issues in the reproduction field, a characteristic of the post-60s. Compared with their American counterparts, the French elites have more leverage to handle social contradictions at the national level and to control the local level. But open ideological divisions within French society and changes in values allied to more permissive structures encourage the state to integrate and accommodate the new demands. Innovation seems to be the trend followed by the newly elected Socialist elites. In contrast, austerity policies, prevailing conservatism, and the absence of a political alternative in the United States seem to indicate that attempts at social change emanating from grass-roots movements are likely to be less favorably met than repressed or neutralized. The example in Europe of new socialist affirmatives and of a humanistic model of society closer to the needs of the people, however, may facilitate and stimulate the transition from social revolts to political projects in America (Castells, 1980: 263). Occurrences in the past have given evidence of the extraordinary potential of the American democracy to generate new dynamic forces of opposition capable of fostering social change.

NOTES

1. The term "state" encompasses the elected government (at all levels) and the administrative bureaucracies, including the domestic police. State power refers to the capacity to make decisions — whatever the nature of the decision — and to enforce them with recourse to some type of sanction to encourage compliance.

2. Such difficulty was encountered recently in New York, when the debate about the reform of Metropolitan Transit Authority (MTA) financing took place. As an editorial in the New York *Times* pointed out, "Whose bus and subway system is it? Mayor Koch says conditions stink but it isn't his system. Governor Carey seems to be occupied, or preoccupied elsewhere. The Assembly Speaker . . . and the Senate Majority Leader . . . looked in vain for leadership elsewhere to support their efforts to put together a fair transit tax package. What they have found instead is an epidemic of political buck-passing and running for cover. . . . Such toe-dancing by the Mayor and others bodes ill for those who actually ride the intolerably decaying buses and subways" (July 4, 1981).

3. In France, firms cannot move away according to their will. Control of industrial siting decisions is established by the responsible governmental administrations and private economic initiatives have to conform with the public scheme of economic development. This is done through a bargaining process (*"concertation"*) between the managers of big business and the managers of the state, in a context of mutual agreement on expansion.

4. This situation is bound to be modified by the new decentralization law.

5. The workers at the *Joint Francais* went on strike in 1972 to demand, among other items, wages aligned on those earned by Parisian workers for the same type of work.

6. One million users from all social classes are carried by the public transit system in Paris, where the housing density is twice that of Manhattan and Brooklyn together.

7. Longer commuting trips between working places and residences, the rigidity of working hours, car congestion, slow bus traffic, and the absence of fare increase from 1960 to 1967 — all heightened costs and deficits in the public transit system.

8. Simultaneously, trade union confederations had thousands of petitions signed and sent to the Ministry of Labor.

9. Demonstrations were accompanied by huge leaflet distributions in the subway and at bus stops, and by information campaigns.

10. American car owners spend more on cars than any western counterparts. The configuration of the cities and the lower density of housing may explain why the car is given priority over public transportation. But other causes are found in federal policies which produced more car owners. The latter became increasingly reluctant to have their taxes subsidize inner-city mass transit systems.

11. The acrimony between transit employees and riders continues to block a political coalition which could address long range transit needs. Such a coalition is possible, as demonstrated by the Italian case (Yago, 1981: 15).

12. For comparative purposes, we have to oversimplify complex processes linked to each project, or even to enclaves within each project. There are obvious differences between Sarcelles, the first public project to the developed in the Parisian region, and Val d'Yerres, the last one. The differences are relative to populations, structures or organizations, and prevailing ideologies. These differences are reflected in the nature and organization of the strikes. The processes are extensively developed in Castells et al. (1978: ch. 5).

13. Only 15 percent of housing was public and 34 percent subsidized between 1959 and 1975 (Ligen and Subileau, 1978: 155). Even so, only 30 percent of new public housing units (HLM) is given to families in priority on the waiting lists. The rest is distributed to state functionaries and the staff of organizations which participated in the financing of the construction. This phenomenon explains why certain HLM units which are highly coveted in Paris may remain vacant for several months.

14. Urban movements are not self-defined in terms of class. It is only for the purpose of clarity that our typology refers to class distinctions.

15. Among nation-states, France has controlled its unity in a rather exceptional way (Gendrot, 1981: 558).

16. The demagogic attempt of the Communist party to capitalize on workers' racism backfired in the presidential campaign of 1981. It endorsed the bulldozing of an immigrant workers' hostel by a Communist mayor in the suburbs of Paris, and it linked an affair of drugs to a Maghrebin family in a vindictive denunciation.

17. The GAMs wanted to have control of the cities' management in order to meet local demands concerning the distribution of space and services. They ran for election and their failure demonstrates the difficulty of running against the traditional partisan alignments. Moreover, after the left was reorganized in 1974, many militants left the GAM and were absorbed by the Socialist party (see Sellier, 1977).

18. It was pointed out that the public service fulfills a fundamental function in regulating regional uneven development and that it therefore benefits the state and employers (Heidenheimer et al., 1975: 171).

19. This pass, called *carte orange*, was not exactly what the users had wanted. Its price, which varied according to distance, disadvantaged working-class populations required to live in remote suburbs because of the price of land in Paris.

20. Linkages between RER (regional railway system) and the subway and buses were established, lines were extended to new suburbs. The creation of bus lines, bus lanes, and shelters made bus travel less difficult.

REFERENCES

ANGOTTI, T. (1981) "The strategic questions for the housing movement: racism, displacement and eviction." Presented at the conference "New Perspectives on Urban Political Economy," American University, Washington, D.C.

BERGER, S. (1977) "Bretons and Jacobins: reflections on French regional ethnicity," in M. Esman (ed.) Ethnic Conflict in the Western World. (Ithica, NY: Cornell University Press.

BOYTE, H. (1979) "Citizen activists in the public interest." Social Policy (November/December): 3-15.

CASTELLS, M. (1980) The Economic Crisis and American Society. Princeton, NJ: Princeton Univ. Press.

———— E. CHERKI, F. GODARD, D. MEHL (1978) Crise du Logement et mouvements sociaux urbains. Paris: Mouton.

CHERKI, E. and D. MEHL (1979) Les nouveaux embarras de Paris. Paris: Maspero.

COHEN, S. and C. GOLDFINGER (1975) "From permacrisis to real crisis in French social security: the limits to normal politics," in L. Lindberg et al. (eds.) Stress and Contradiction in Modern Capitalism. Lexington, MA: D.C. Heath.

ETZIONI, A. (1968) The Active Society. New York: Free Press.

FAINSTEIN, N. and S. FAINSTEIN (1980) "Bureaucratic enfranchisement and the politics of neighborhood development." Presented at meetings of the American Political Science Association, Washington, D.C.

GENDROT, S. (1981) "La Société française: divisions et consensus. French Review 4 (March): 558-565.

———— (1980) "New York, la démocratie prise au piège." Etudes (November): 465-478.

———— and J. TURNER (1982) "Ethnicity and class: politics on Manhattan's lower east side." Ethnic Groups.

GITTELL, M. (1981) "Violence as a political strategy: defending the status quo." Presented at meetings of the American Political Science Association, New York, N.Y.

———— (1980) Limits to Citizen Participation. Beverly Hills: Sage.

GODARD, F. et al. (1973) "La rénovation urbaine à Paris. Paris: Mouton.

GREMION, P. (1978) "Les associations et le pouvoir local." Esprit 6.

HARTMAN, C. (1980) "Transportation users' movements in Paris in the 1970's." Presented at a symposium on "Citizen Participation in Western Europe" sponsored by the German Marshall Fund of the United States, Washington, D.C.

HEIDENHEIMER, A., H. HECLO, and C. ADAMS (1975) Comparative Public Policy. New York: St. Martin's.

JONES, D. (1979) "Not in my community: the neighborhood movement and institutionalized racism." Social Policy (September/October): 44-46.

LIGEN, P. Y. and J. SUBILEAU (1978) "The new development policy for the city of Paris," in G. Wynne (ed.) Survival Strategies. Paris and New York. New Brunswick, NJ: Transaction Books.

MEHL, D. (1980) "Les voies de la contestation urbaine." Annales de la Recherche Urbaine.

NEWFIELD, J. and P. DUBRUL (1977) The Abuse of Power. New York: Viking Press.

PIVEN, F. and R. A. CLOWARD (1979) Poor People's Movements. New York: Vintage.

SCHATTSCHNEIDER, E. E. (1960) The Semi-Sovereign People. Hinsdale, IL: Dryden.

SELLIER, M. (1977) "Des groupes d'action municipale." Sociologie du Travail 1.

STONE, C. (1980) "Social stratification and community power: decisions, non-decisions and beyond." Presented at meetings of the Political Studies Association, Manchester, England.

SMITH, D. H. et al. [eds.] (1972) Voluntary Action Research: 1972. Lexington, Lexington Books.

WOLFE, A. (1978) The Seamy Side of Democracy. New York: Longman.

YAGO, G. (1981) "The Coming Crisis of U.S. Transportation." Presented at American University, Washington, D.C.

YATES, D. (1977) The Ungovernable City. Cambridge, MA: MIT Press.

About the Contributors

PAUL ADAMS was educated at University College, Oxford, the London School of Economics, the University of Sussex, and the University of California, Berkeley. He taught for four years at the University of Texas at Austin and is currently Associate Professor in the School of Social Work, University of Iowa. His book, *Health of the State*, will be published by Praeger in 1982. He is engaged, in collaboration with Gary Freeman, in the comparative study of social security and social services in Europe and North America.

MANUEL CASTELLS is Professor of City and Regional Planning at the University of California, Berkeley. Previously he was Professor of Sociology at the Ecole des Hautes Etudes en Sciences Sociales, Paris. He has published twelve books in several languages, including *The Urban Question* (MIT Press, 1977); *City, Class and Power* (Macmillan, 1978); *The Economic Crisis and American Society* (Princeton University Press, 1980); and *Citizens: A Cross-Cultural Theory of Urban Social Movements* (Edward Arnold Press and University of California Press, 1982). He is coeditor of the *International Journal of Urban and Regional Research*.

PAOLO CECCARELLI is Professor of Urban Analysis in the Department of Social and Economic Analysis of the Territory, University of Venice, Italy. He has been Visiting Professor at the Massachusetts Institute of Technology and the University of California, Santa Cruz. He is currently examining the social and political consequences of deindustrialization of large urban areas in Europe and the United States. He is involved in a comparative study of Detroit and Turin sponsored by the Center for European Studies, Harvard University.

SALAH EL-SHAKHS is Professor of Urban Planning and Director, School of Urban and Regional Policy at Rutgers University. He is a frequent consultant to international planning firms and organizations. His most recent publications include a United Nations Conference Document on *National and Regional Issues and Policies in Facing the Challenge of the Urban Future* (1980) and a coedited volume on *Development of Urban Systems in Africa* (1979).

NORMAN I. FAINSTEIN is Professor of Urban Affairs and Policy Analysis, Graduate School of Management and Urban Professions, New School for Social Research. He and Susan Fainstein are completing a study of the comparative politics of development in U.S. cities. They are the authors of *Urban Political Movements*, editors of *The View From Below*, and have published articles on urban theory, planning, public policy, and local politics. They are American coeditors of the journal, *Ethnic and Racial Studies*.

SUSAN S. FAINSTEIN is Professor and Chairperson of the Department of Urban Planning and Policy Development at Rutgers University — the State University of New Jersey. She and Norman Fainstein have collaborated on a number of projects, as indicated above.

GARY FREEMAN teaches in the Government Department at the University of Texas at Austin, where he specializes in comparative public policy and American and European politics. He is the author of *Immigrant Labor and Racial Conflict in Industrialized Societies: The French and British Experience, 1945-1975* and articles on social policy in Europe and America. He is currently engaged in a comparative study of the relationship between the present economic crisis and the development and reform of old-age pension systems in advanced capitalist states.

SOPHIE N. GENDROT teaches at the Institut d'Etudes Politiques, Paris, where she is doing research on neighborhood conflicts and local power in the United States. A Tocqueville Scholar, she has been a visiting scholar at Columbia University and New York University, and she has published several articles in French and American journals.

KATHLEEN GILLE is a Ph.D. candidate in the Department of Political Science at the University of Chicago. Her present research interests concern labor and politics in the contemporary United States. She is currently employed as Special Projects Assistant for Congressman David Bonior.

IAN GOUGH joined the Department of Social Administration at Manchester University in 1964, where he lectures in social policy and political economy. He is a contributor to *New Left Review*, and is now on the editorial board of *Critical Social Policy*. His book, *The Political Economy of the Welfare State*, published in 1979, will shortly be appearing in Spanish and Swedish translations. In 1978-1979 he was visiting professor of economics at the University of California, Berkeley. He has also lectured in Australia, Italy, Spain, and Denmark.

IRA KATZNELSON is Professor and Chairman of the Department of Political Science at the University of Chicago. His most recent book is *City Trenches: Urban Politics and the Patterning of Class in the United States*. He is also author of *Black Men, White*

Cities and co-author of *The Politics of Power: A Critical Introduction to American Government.*

PETER MARCUSE (J.D. Yale Law School, Ph.D. Berkeley) teaches urban planning at Columbia. He has been Chair of the City Planning Commission of Los Angeles, and is a member of Community Board 9 in Manhattan. He is author of "The Myth of the Benevolent State" and other pieces on housing and planning policy. The present chapter is part of ongoing comparative work on the determinants of housing policy; comments and criticism would be welcomed.

S. M. MILLER is Professor of Economics and Sociology at Boston University. He is the co-author of *The Future of Inequality* and is currently working on a comparative study of European poverty programs. He has recently completed a manuscript with Donald Tomaskovic-Devey on *The Recapitalization of Capitalism.*

KAREN MURPHY, MCP, is currently Liaison Officer at the Mayor's Office, City of New York.

RICHARD C. RICH is Associate Professor of Political Science at Virginia Polytechnic Institute and State University. His articles on urban neighborhoods and citizen participation have appeared in *American Behavioral Scientist, Urban Affairs Quarterly, American Journal of Political Science, Urban Studies,* and *Social Science Quarterly.* He is Deputy Editor of the *Journal of Community Action,* coeditor of a series of books on urban public policy published by the State University of New York Press, and Associate Editor of the *Journal of Applied Behavioral Science.*

DONALD TOMASKOVIC-DEVEY is a research assistant in the Department of Sociology at Boston University. He is currently working on a dissertation about structural change in the U.S. division of labor and has recently finished a manuscript with S. M. Miller on *The Recapitalization of Capitalism.*

JOHN WALTON is Professor and Chair of the Department of Sociology, University of California, Davis. He has done research and published in the areas of the political economy of underdevelopment (e.g., *Elites and Economic Development)* and comparative urbanization (e.g., *Urban Latin America* with Alejandro Portes and *The City in Comparative Perspective* with Louis Masotti). In 1981, with Alejandro Portes, he published *Labor, Class, and the International System.* He is currently working on a comparative study of revolution and underdevelopment.

MARGARET WEIR is a Ph.D. candidate in the Department of Political Science at the University of Chicago. She is currently studying employment policies in the United States and Sweden in the 1930s.